Dawn

Dawn

The Origins of Language and the Modern Human Mind

Rik Smits

Routledge
Taylor & Francis Group
LONDON AND NEW YORK

First published 2016 by Transaction Publishers

Published 2017 by Routledge
2 Park Square, Milton Park, Abingdon, Oxon OX14 4RN
711 Third Avenue, New York, NY 10017, USA

Routledge is an imprint of the Taylor & Francis Group, an informa business

Nederlands letterenfonds
dutch foundation for literature

The publisher gratefully acknowledges the support of the Dutch Foundation for Literature.

Library of Congress Catalog Number: 2015042434

Library of Congress Cataloging-in-Publication Data

Names: Smits, Rik, 1953- author.
Title: Dawn : the origins of language and the modern human mind / Rik Smits.
Other titles: Dageraad. English
Description: New Brunswick : Transaction Publishers, [2016] | "In this work, originally published as Daferaad, in Dutch, Rik Smits theorizes that language could not have developed originally as a system of communication." | Includes bibliographical references and index.
Identifiers: LCCN 2015042434 (print) | LCCN 2016000129 (ebook) | ISBN 9781412862578 | ISBN 9781412862103 (eBook) | ISBN 9781412862103 ()
Subjects: LCSH: Language and languages--Origin. | Language and languages. | Human evolution.
Classification: LCC P116 .S6313 2016 (print) | LCC P116 (ebook) | DDC 401--dc23
LC record available at http://lccn.loc.gov/2015042434

ISBN 13: 978-1-4128-6257-8 (hbk)
ISBN 13: 978-1-4128-6265-3 (pbk)

Contents

Prologue		vii
1	Night	1
2	The Animal Inside Us	29
3	Goodbye to the Animal Kingdom	55
4	Phantom Limbs and Figments of the Mind	81
5	Shamans in the Shadows	105
6	Voices on the Plains	137
7	Dawn	165
Epilogue		191
Bibliography		207
Index		213

Prologue

The Forbidden Field

Morgan Kavanagh was a man on a mission to whom nobody would listen. Yet he had solved one of the greatest mysteries in human history, or so he thought. Buried deep within ancient myths, he had discovered the divine origins of human language and thought.

In 1844, he published his findings in a hefty tome. He then spent the next thirty years frantically seeking recognition of his work, but to no avail. Seven times he competed for the prestigious Volney Prize for linguistics, but never once was he awarded those wretched 1,200 francs. Worse, his voluminous submissions, some in French, others in English, soon ceased to be taken seriously.

Kavanagh was first disregarded, then forgotten, like an itch to be ignored away. This had a lot to do with the founding in 1864 of a select company of linguists, the *Societé de Linguistique* in Paris. The Societé was not meant to be a place for small talk or idle philosophical speculation. It was to be a venue for serious scholars, the kind of association to which everyone who was anyone in the world of linguistics would want to belong. It would be devoted exclusively to the study of language, legend, traditions, customs, and other matters of ethnographic interest. At least, this was what it promised in the first of its articles of association, passed in 1866. But it was the now notorious Article 2 that gave voice to the main concern of the founders: "The Societé shall not accept any contribution concerning the origin of language or the creation of a universal language."

It seemed a strange, illogical ban. Less than seven years after Charles Darwin had published his revolutionary *On the Origin of Species*, this brand new association declared out of bounds the origins of the very subject it set out to explore. The impact of Darwin´s book, which turned the spotlight on evolution as the driving force behind all life on earth, had been felt far and wide. It had given new impetus and focus to semievolutionary ideas that had been floating around for more than a century among naturalists, as scientists were called in those days. And it had finally freed people from the time-honored fetters of faith. At long last, one could publicly dispense with paying lip service to the biblical story of Adam and Eve and the six days of creation, some six thousand years earlier. *On the Origin of Species* had sent into turmoil all thinking on how man and beast and their defining characteristics had originated. So how could the Paris Societé slam the door on the question of origins at that very moment? And why did its London counterpart follow suit in 1872?

The reason was the very climate of tempestuous debate that had followed in Darwin's wake. Suddenly the world abounded with fascinating mysteries, but there were no reliable data or valid methods to solve them. That deficiency was readily made up for by the optimism and enthusiasm of both professional scholars and often quite colorful amateurs. The temptation to speculate to their hearts' content was all too great to resist, not least among those who tackled the most essentially human quality of all: language.

This temptation was understandable. People trying to unravel the ontogeny of our purely physical properties could turn to an ever-growing number of fossils, but the explorers seeking the sources of human language soon found themselves wading through quicksand. Language simply does not solidify: it involves only highly perishable tissue, like brains, small muscles, and cartilage. What is more, it would be many years before Karl Popper would force scientific practice into a rigid methodological corset, insisting on transparency and verifiability of method. This made it even harder to separate the wheat from the chaff, and meant that much so-called research into the origins of language comprised little more than religious ravings, naive delusions, and sometimes highly complicated fantasies. There were Kavanaghs everywhere, each with a theory more sprawling and ill-founded than the next.

When it came to the age-old idea of a universal, "ideal" language, the situation was just as dire. In biblical terms, this would be the language that people spoke before the Tower of Babel was built. The desire to rediscover it was nearly as old as the mythical, primeval language itself. The world's first known historian, Herod, reported on an experiment conducted by pharaoh Psamtic I as early as 650 BCE. Psamtic ordered that two newborn babies be brought up in isolation, with no human contact except each other. The clever king reasoned that this would leave them no option but to develop a language of their own, which would logically be the coveted original language. And, as Herod has it, he turned out to be right. When the infants began babbling, the first thing that came out sounded like *bekos*. Psamtic's scholarly advisers recognized this as the Phrygian word for "bread," and concluded that the first language had to have been Phrygian.

History has seen many more such crude attempts at discovering the primordial language, experiments that all ended badly. The children would die, whereupon the instigator would lose interest, or vice versa. But that did not stop all kinds of dreamers, fantasizers, and amateur linguists from concocting their own holy grail of historical linguistics—men like the sixteenth-century luminary Goropius Becanus, for example. Becanus was a sensible man who rose to the exalted positions of personal physician to King Philip II of Spain and city surgeon of Antwerp. But after that he somehow lost his way, taking it upon himself to prove that Adam and Eve had spoken Flemish in the Garden of Eden. That placed Paradise firmly in Flanders, which just happened to include Becanus's own native village of Gorp, nowadays part of the Netherlands.

Just as numerous were those who endeavored not to find the long-lost first language, but to create the perfect language themselves. These were idealists, hoping that a single world language would unite all mankind. Rare exceptions aside, such as Ludwik Zamenhoff, the Pole who invented Esperanto, they and their languages have all been forgotten.

Upstanding nineteenth-century linguists abhorred such activities. In 1786, the former chief justice of Bengal, Sir William Jones, had already shown that languages like English, Greek, and the extinct Indian language Sanskrit must have sprouted from a single source. Despite their obvious differences, in crucial respects they accorded with one another in deeply meaningful ways that simply could not be accidental. Languages, it transpired, were organized in families, and once that idea had taken root, professional linguists devoted themselves first and foremost to the construction of family trees, in particular the pedigree of Jones's so-called Indo-European family. Their vision of a primeval language was the ancient language that had fathered all known Indo-European languages, and their object was to reconstruct it. They called it Proto-Indo-European.

These linguists were bent on reasoning back through time as far as the available written sources, logic, and, if all else failed, conjecture would let them. They were systematic workers who loathed the romantically or religiously inspired figments of the imagination of the likes of Kavanagh and Becanus. It was this despair at the unstoppable avalanche of dreamers and zealots that drove the founders of the Societé to place a taboo on the whole subject of the origins of language. And it worked. For more than a century, any linguist who wanted to be taken seriously would carefully avoid anything to do with how, where, and when human language had originated.

But that was then. The world has since come a long way, and so has our knowledge of the human language faculty. Since about 1985, interest in the question of how we came by it has slowly rekindled. It is still a small flame, but it is bright enough to shed a flicker of light on the matter of the emergence of human language, and with that on the dawn of our existence as modern humans. It is a story that begins in foggy, dusty twilight, more than three million years ago.

1

Night

Through the clouds of scorching dust and ashes that Sadiman has been belching up for days, a small family treks across the grey, bleak savannah. They pay little heed to the rumbling volcano, whose glow paints the dust-filled sky an ominous orange. They have become used to it.

The man leads the way. The woman follows, a child clinging to her left hand. The girl has trouble keeping up, in part because despite her parents' agitation she is playing a game, trying to tread precisely in her father's footsteps. Every now and then, the booming of the volcano is punctuated by pelting showers and crashing thunder, the rain turning the thin blanket of ash that covers everything into slick mud.

At times, they catch a fleeting glimpse of other creatures in the dusty half-light. Buffalo, antelopes, the occasional large predator. They ignore them; they have other worries. They walk fast, in silence, panting with exertion. When the child trips, the woman yanks her upright. "Keep moving, dammit," she snaps, without breaking her stride. And then, soothingly, "Just for a little while longer."

Nobody knows whether this little family made it to wherever they were going in the dust bowl Sadiman was turning their part of East Africa into. Today, only their footprints remain—footprints that solidified in the mud and were buried under new coats of ashes and dust, only to be stumbled upon over 3.5 million years later by Andrew Hill, a paleontologist working the Tanzanian site of Laetoli. That was in 1976. What exactly was going on when the footprints were made—something like the scene above, or something altogether different—we can never be sure. But one thing is clear: it is extremely unlikely that whoever walked there could actually talk or was capable of anything even remotely like it.

They were very different from us, these wanderers. They were small and ape-like, with brains weighing less than a pound, much like those of modern chimpanzees. Most experts attribute the Laetoli footprints to creatures like Lucy, the famous *Australopithecus afarensis* skeleton that had been unearthed nearby only two years before Hill made his discovery. "Lucy" herself, however, presumably lived nearly half a million years later.

Lucy and her kind walked upright. That much is certain. But although they are considered our more or less immediate ancestors, these creatures were nowhere near human. The oldest known tools—mere shards that were intentionally struck off larger stones—date back approximately 1.75 million years. That is more than a million years after Lucy's kind disappeared from the face of the earth. Meanwhile, one of the most mysterious developments in human history took place: the unprecedented enlargement of our neocortex, which we have to thank for both our oversized heads and the exceptional faculties that make us human.

A million years—just think about the scale of things here. If the distance separating us from Julius Caesar is one step on an ordinary staircase, a million years would be as high as St. Stephen's tower in Westminster in London, which houses Big Ben. Today, women in Western Europe on average give birth for the first time when they are around thirty years old. At that rate, a million years equals 33,333 generations, whereas we are only sixty-seven generations away from Caesar, fewer than two hundred from the invention of writing and roughly three hundred from the emergence of agriculture and the founding of the earliest permanent settlements. Such a vast expanse of time is in itself good reason to deny our distant ancestors the capacity for language. One simply cannot imagine a species having such a powerful instrument at its disposal for so many generations without ever using it in any recognizable way: never accumulating and pooling knowledge, never building up anything resembling a culture, barely showing any sign of creativity. How likely is it that creatures that possessed the ability to speak would have remained unable for so long to come up with anything to say? They would literally have bored themselves to death—in which case there would have been no "us."

* * *

Yet here we are. So it makes good sense to ask when and why our language faculty did come into existence. For whichever way we look at things, human language remains an astonishing, even miraculous phenomenon. We experience it as our most human, most defining characteristic. It is a phenomenon that distinguishes us fundamentally from all other forms of life. We are prone to attributing virtually any human trait to a pet. Most dog owners bestow the power to think, reason and understand people on their four-legged favorites as a matter of course. And people routinely talk to dogs, even other people's dogs, as if they were having a perfectly normal chat with a fellow human. But there is one difference: we never expect animals to talk back, to engage in actual conversation. Not dogs, and not even parrots. Nor do we expect them to be able to communicate in anything other than their usual way, which is why we never blame Fido for answering nothing but "woof!"

Language is characteristically ours. It is what we *are*—or at least, that is how we feel. If there is any truth to that feeling, if language is indeed uniquely human, then at some point in time something must have happened that sent our forebears off on a new, essentially different track—something that caused a deep rift between us and all the rest of the animal kingdom.

Does this rift really exist? Does our language faculty set us fundamentally apart from our fellow creatures? Or do humans just possess more of the same stuff other species have? These questions have been subjected to much scrutiny over the past hundred years or so, and from time to time the matter of the origins of human languages has stealthily reared its head again, in spite of the 1866 ban on the subject by the Societé de Linguistique.

Roughly, the matter of the apparent uniqueness of human language has been broached from two perspectives. The first is that of current mainstream linguistics, generative theory, developed by the American linguist Noam Chomsky in the mid-1950s. Generative linguists consider the human language faculty to be unique. These "Chomskyans" are opposed by a colorful coalition of ethologists, psychologists, linguists, and would-be linguists of all persuasions, who entertain altogether different notions. Most flamboyant among them are those who have left no stone unturned in their efforts to teach a wide variety of animals something resembling human language. Animals like Alex, the miracle parrot who died in 2007; Koko, the gorilla who failed miserably at online chat; and the chimpanzee Nim, flippantly nicknamed Nim Chimpsky. The "Chimpskyans" believe that there is no principled difference between man and beast, and so—given proper training of course—animals should be able to learn to use language.

To Chomsky and his colleagues, such ideas have always been anathema. Right from the outset Chomsky maintained that only humans

Noam Chomsky at 84 in 2013 (Photographed by Oliver Abraham).

have a language faculty. That faculty, he said, was based in a special language organ in the brain, which was there exclusively for the production and reception of language. It all sounded a bit dogmatic, conceited even. Storms of protest arose, and angry incomprehension ran rife, for there was no trace of a dedicated language organ to be found. On the contrary: as the years went by and more and more areas of the brain were identified that were involved in language processing, each and every one of them proved to serve other purposes as well.

Had the great Chomsky really been so utterly wrong? Not really. While the concept of a language organ was in one way quite essential to him, from a different perspective it was small fry. Chomsky is a theoretician. The description of our language faculty that he has strived all his life to achieve is not a picture in a biological atlas, but an abstract model. To him, the language faculty is not a gooey clump of cells, but a system of strictly logical rules that produces the same effect as a "real" language faculty. Such a model has much in common with flight simulation software. It looks nothing like actual airspace, full of wind, rain and up-currents, but for the trainee pilot it replicates all the vicissitudes of actual flight in full.

Claiming that some specific, isolated corner of the brain was dedicated to language and language alone simply enabled Chomsky to put all other psychological traits aside and concentrate exclusively on the language faculty proper. That way, the enterprise that he set so much store by remained manageable. What mattered first was to construct a solid theory of how everyday words and sentences were structured. That alone would be quite an achievement. Once that framework existed, it would be possible to delve into the relationship between language and other properties of the mind, like memory, thought, and emotions. And not until then would it make sense to start thinking about how the whole thing was embodied in the world of flesh and blood.

Apart from such practical, strategic considerations, there was also a principled point to be made. Chomsky wished to stress that language is part of our biological endowment. It was not, as the prevailing wisdom had it when he set out on his quest, a purely cultural phenomenon on a par with knitting, football or playing the piano. Those are all skills that a child must acquire step by step, whereas language, Chomsky wanted to make clear, is something with which we are born. It is as much a part of our bodies as eyesight or the ability to walk. And like eyesight and walking, language is among the biological skills that are not yet fully developed at birth. They reach maturity only during childhood, and require input from the environment to do so. Eyes need light, walking depends on gravity and the maturation of the human language faculty requires that we hear talk in what is to become our mother tongue.

Thus Chomsky branded language as being rooted in *nature* rather than *nurture*, as something more to do with hereditary factors than with the social environment, innate rather than acquired. Input from the environment played a relatively superficial role, he maintained. To openly take and defend such a position was not easy in the intellectual climate of the early 1960s, especially in America.

* * *

The mid-twentieth century was the heyday of behaviorism, a school of thought that had held psychology and the humanities in America in a stifling grip since the Roaring Twenties. It was a doctrine that, in its most radical form, denied the existence of virtually all spiritual life and shunned all things purportedly hereditary and innate. To behaviorists, all behavior was the logical outcome of external impulses, as predictable as billiard balls obeying the impact of the cue. This held for man and beast alike. All were born with minds as empty as the proverbial tabula rasa, a clean slate. From these humble beginnings, infants gradually evolved into complex adults through a process of habit formation. All they needed, they picked up from the environment, under the pressures of satisfaction and want, pleasure and pain. In the eyes of a behaviorist, both people's characters and whatever steered their actions were as plastic as wet clay, to be molded at will.

The intellectual father of behaviorism was John B. Watson, an American born in 1878 into a broken, destitute family in a South Carolina backwater. Despite these inauspicious beginnings, he climbed to the rarefied altitudes of a full professorship at Johns Hopkins University. And that was only for starters. Just as he was really hitting his stride, he fell from grace, amid much public fanfare, over an affair with a graduate student that has remained shrouded in mystery to this day. Both his position and his reputation went out the window. But somehow, Watson managed to pull himself out of the mire by his own hair to shine once more, this time as vice president of J. Walther Thompson, one of the most prominent advertising agencies of the day.

Watson's career seemed to be living proof of his own claim that he could mold a highly skilled specialist from any healthy infant, regardless of its race and social background. It is also quite understandable that his ideas found so much acclaim, especially in the United States. Watson and his theory epitomized the American dream, the belief that every newspaper boy can become a millionaire, if only he tries hard enough.

By the time Chomsky was in his twenties, behaviorism had spread from psychologists to much wider circles. The conviction that the human mind could be shaped and reshaped at will had achieved almost divine status in the aftermath of World War II. After all, had the Nazis, with their *Übermensch* and their racial theories, not proven beyond any shadow of a doubt what horrendous cruelty could be spawned by heredity-based ideologies? Behaviorism was more than just an alternative. After all the bloodshed, it also offered consolation, hope and reassurance. If all behavior was acquired by learning, then the atrocities of the Germans and their collaborators were not inevitably in our blood. They had simply been misguided, brought up the wrong way, their minds warped. Likewise, to a behaviorist the horrors that the Allies had visited upon their foes in return were but responses to signals received from the Nazis. People could breathe a sigh of relief, safe in the knowledge that none of this had any bearing on someone's true character.

* * *

Chomsky's conviction that language was a unique, innate property of mankind, residing in a dedicated mental organ, vehemently clashed with the received ideas underpinning the thinking of the Chimpskyans.[1] Most prominent among them were the many psychologists and psycholinguists who in the course of the twentieth century attempted to train gorillas, chimpanzees, bonobos, and other animals to use language. Their jumping-off point was the idea that there were no principled differences between mankind and its closest relatives, apes, and other primates. Some among them were card-carrying behaviorists who abhorred even the slightest hint that something innate might underlie sophisticated behavior, including the use of language. They were convinced that it was only a matter of finding an appropriate conduit, a suitable mode of communication, for an animal to be able to express itself. Once such a conduit were identified, systematic training would produce animals capable of using language. Other Chimpskyans held different, or at least less extreme, views. To them, the tenet that there were no fundamental differences between man and animal was first and foremost of practical importance. It was a useful null hypothesis, a point of departure for their research. If it was wrong, this would become clear when it proved impossible to teach animals to use language.

The adventures of the ape trainers make for fascinating, if only too often dramatic and tragic stories. Take Washoe, who died in 2007. From 1966, when she was in her infancy, until 1970, this chimp was a full member of the Gardner family of Reno, Nevada. The couple, both psychologists, raised her exactly as they would a real child. Washoe ate at their table, wore nappies and slept in a child's bed. The idea was that she would learn American Sign Language, the sign language of the American deaf. Other apes, like Nim, were kept in purpose-built facilities and trained by an elaborate staff of specialists. But invariably, once they grew past childhood, the animals would become too dangerous to handle, and the project would have to be abandoned. Or, equally invariably, the money would run out. Usually the animals ended up in a lonesome cage, forgotten by their researchers and too domesticated to be released into the wild or become a member of a captive group.

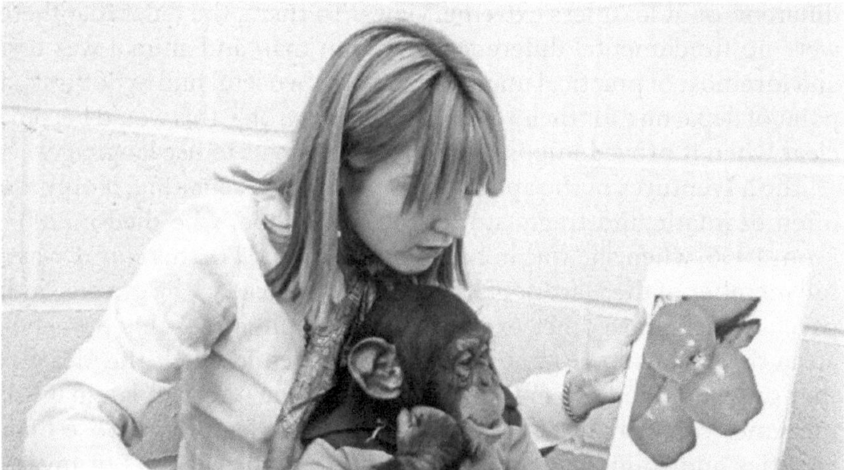

Nim with his favorite caretaker, Laura-Ann Petitto, around 1976 (Copyright Dr. Laura-Ann Petitto, University of Toronto).

Despite all the tall tales to the contrary, this endless series of experiments consistently failed to produce the desired result: a lingual ape. Speech was obviously out of the question, for biomechanical reasons. Apes' mouths and throats differ from those of humans, making it impossible for them to produce many of the strings of sounds needed for speech. But such shortcomings could be circumvented. Sign language was one option, another was a kind of keyboard with symbols representing objects and actions. Indeed, some apes learned to use symbols to represent foods, frolicking and other desirables of ape life. But that was it. Not a single reliable instance has ever been recorded of an ape producing even the simplest, most rudimentary sentence out of symbols, signs or sounds, or in any other way.

In fact, the "talking" apes bore no resemblance to human language users whatsoever. For one thing, with one or two rare exceptions they showed fairly little interest in the matter. They needed a strong hand to keep them on track, and rarely, if ever, learned anything of their own accord. Their achievements, such as they were, were the painstaking fruits of relentless training. As soon as the trainers left off, progress would not only grind to a halt but most of the apes would stop using the tricks they had acquired altogether. Nothing could be farther from the way normal human children acquire language, with no coaxing at all. Human children cannot *not* acquire language, and they will keep on talking all their lives. If no one else is around, we will jabber at the dog, chat with the cat, yell at the television and talk to ourselves, be it out loud or silently. All children acquire and use language, and they all do so with no effort whatsoever, regardless of whether they receive explicit instruction or not. Even the scruffiest ten-year-old urchin can put in his two cents with the best of them, thank you very much, although he may not be overly polite or eloquent.

Yet another difference was that no ape ever started naming things the way children do. Moreover, they rarely initiated a "conversation," and their repertoire never really developed beyond set phrases like "give me this" or "I want that." They added nothing of their own to what they were taught, and they did not experiment. In short, the whole effort made it eminently clear that, although apes could be pressured into learning and remembering quite a lot, they never got past the aping stage, so to speak, performing tricks that had nothing to do with actual human language.

Quite apart from these experiments with great apes, there was also a steady stream of less bizarre research into the true character of animals. People like Niko Tinbergen of the Netherlands and the Austrian Konrad Lorenz pioneered the field of ethology, the study of the behavior of animals in their natural environment. More and more others followed, scrutinizing more and more different species. Whether these researchers like it or not, the public at large has always understood their work to be an inquiry into how *human-like* animals behave. This goes some way to explaining why the ethologist Frans de Waal's books, for instance, meet with such excitement, occasionally spawning such far-fetched ideas as "the bonobo as a business model."

This type of research had two major consequences. On the one hand, there was growing awareness that many traits that had always been considered typically human also occurred in animals, at least to some extent. On the other, out of sheer enthusiasm animals were (and are) ridiculously overestimated.

A prime example of such overestimation appeared in the August 2007 issue of the journal *Ethology*. Female macaques make a specific sound, known as a *girney*, which they direct at mothers with children, and also at children in the absence of their mothers, but never at their own offspring. For this reason, it had long been thought that the sound was aimed first and foremost at fellow mothers. But now, researchers had discovered that the girney was primarily meant for the young after all, and, moreover, that it carried no meaning, not even to signal friendly or less friendly intentions on the part of the utterer. It served purely to attract and keep the attention of the macaque young, who, incidentally, paid little attention to it. Perhaps, or so the writers claimed, this might resemble motherese: the language people use when they communicate with babies, with an abnormally high pitch and exaggerated voice inflection.

This idea evidently holds no water. Where motherese is found—and there are many cultures in which babies are not spoken to at all—it is used primarily to converse with one's own child. It consists of much more than just a high-pitched melody, and, unlike the macaque young, babies tend to respond to it with great enthusiasm. That, after all, is precisely the point of motherese: people keep at it until the baby starts smiling, spluttering and babbling back. These distinctions notwithstanding, the press came up with spectacular headings like "Monkey Motherese." This was pure nonsense, which merely strengthened the mistaken idea that monkeys possess language.

Even more outlandish was the story about dolphins that hit the media in the late spring of 2006: dolphins had and used proper names!

"Dolphins communicate like humans by calling each other by 'name,' BBC News raved on May 8, and quoted biologist Vincent Janik as saying: "it means that these animals have evolved the same abilities as humans."

This story was accurate only in that individual dolphins do use so-called signature whistles to identify themselves to others. In itself this is nothing special; lots of species that live in groups do so. Penguins, for instance, use signature whistles to pick out their young or their parents from among thousands of tightly packed, apparently identical birds. More interesting was Janik's finding that, unlike most other species, bottlenose dolphins did not identify themselves through the unique sound of their voices. Instead, each dolphin's signature whistle has its own characteristic form and melody. But although somewhat unusual, this behavior too is far from unique. The chicks of certain kinds of songbirds, for example, first copy their parents' signature whistles, then develop a variation on them to give them a "personal" twist.

In fact, Janik had claimed nothing about human communication, let alone language. All he did was draw a parallel between the signature whistles of dolphins and the way people use proper names. But even that was neither here nor there, for signature whistles have nothing in common with human names and our urge to name things. For one thing, people do not make up their own names. Instead, usually, both our Christian names and surnames are bestowed upon us by others. And—under ordinary circumstances at least—nobody goes around shouting their own name all the time, whereas half of the utterances made by Janik's dolphins were signature whistles.[2] Human proper names are typically used by *other people* when addressing you, when they try to attract your attention or have a nice chinwag about you when you are not in the room.

The long and the short of it is that what bottlenose dolphins do is no different from the achievements of songbirds, or, for that matter, the average lighthouse. Signature whistles function exactly like the identification code of a buoy at sea or a transponder in an airplane. They neither arise nor function in the same way as human proper names, giving the lie to any notion that these marine mammals are developing anything even remotely resembling human language.

Almost a century of animal research has left us with just one main insight. Most of the traits long considered typically "human" have been shown to occur, however slightly, in other species too—but never anything resembling human language. This means that the animal world holds little promise of shedding light on what exactly our capacity for language is, or how and why it came into being. As a consequence, we must now turn to the few things we know about our own ancestors in these respects. We must consider what little knowledge we have about Cro-Magnons, the very first modern humans, and about hominids, the semi-human species that roamed the earth before them. About, for instance, the largely mythical creature called Black Eve.

* * *

Black Eve first arrived on the scene in 1988. She was the brainchild of two British paleontologists, Chris Stringer and Peter Andrews, and almost immediately became the favorite pet of anyone with an interest in human evolution. Stringer and Andrews claimed that they could reconstruct the complete family tree of modern mankind, all the way back to a single African woman, Black Eve. She was the primordial mother, a creature who had lived somewhere in Central Africa some two hundred thousand years ago. She could be traced, the paleontologists said, by means of mitochondrial DNA, minute specks of DNA that are not part of any chromosome.

As the name suggests, mitochondrial DNA lives inside organelles called mitochondria, infinitesimally small energy factories that are rife in both normal cells and female gametes, but absent from sperm. For this reason, mitochondria and their DNA transfer directly and unchanged from mother to child. Fathers have nothing to do with it. This gave Stringer and Andrews the idea that the number of differences between the mitochondrial DNA of two populations was an exact measure of the genetic distance between them. In fertilization, nuclear DNA from the mother mixes with that of the father to form novel, unique strings of DNA in the child. But mitochondrial DNA is never mixed, so it can only change through random mutation. Moreover, random mutations, it is safe to assume, occur with a certain statistical regularity. Counting the differences in the mitochondrial DNA between two groups of a species should therefore reveal how long ago those groups had become isolated from each other. If a random mutation occurred on average once every hundred years, say, the number of differences would equal the number of centuries that had passed since separation. By means of this genetic clock, we might also be able to figure out what route early humans took when they struck out from Africa, and when—or so Stringer and Andrews claimed.

The result was the Out-of-Africa hypothesis, which claimed that early man spread from Central Africa to the Middle East in two waves.

After the first wave, some 150,000 years ago, our distant ancestors had pushed on from their new territories to colonize the great land masses of Europe and Asia. During the second wave, fifty thousand years ago, they roamed even further, all the way to the farthest corners of New Guinea and Australia.

Although there are serious objections to this model, it became so popular so quickly that even certain linguists dared once more to publicly ponder the enigma of the emergence of language. The 1866 taboo had lost its edge, and its hold over linguists all but vanished in the decades following Eve's appearance.

But old habits die hard. Despite the modest tone of the rekindled debate on the origins of language, it immediately zeroed in once more on the old chestnut of whether language could be found in animals. In other words, are we unique among species? The harder one tries to answer this question, the more elusive it seems to become. Recent years have seen yet another endless stream of experiments, some more sophisticated than others, hinting that animals share certain characteristics traditionally ascribed only to humans: traces of tool making and use, behavior reminiscent of symbolic thought, a smattering of culture, a pinch of logical reasoning, some lying, cheating and manipulation, signs of self-consciousness, theories about what other individuals might know and think, and so on. At the same time, geneticists consistently find the differences between human DNA and that of the great apes to be smaller than expected. This is so much the case that some social scientists and ethicists have attributed almost human forms of consciousness to animals and begun advocating animal rights. Yet, the chasm separating man and animal remains as unfathomably deep as ever. Go to a zoo and this is plain to see: Who is inside the cages and who is not? Who is interestedly ogling whom, and who are doing their level best to ignore the other?

* * *

There is just no denying the deeply rooted differences between man and beast, differences that are easy to outline in broad strokes. Our endlessly flexible and creative culture is an example, as is our capacity for abstract and rational reasoning. The same is true for the very fickleness and unbounded variability of human behavior. Consider, too, the all-embracing consciousness of future and finiteness that makes us worship gods, experience midlife crises and carry the memory of our deceased loved ones and our exes with us for life. And, lest we forget, there is our indomitable urge to learn language and keep on using it all our lives, every waking hour of every day.

That said, it remains difficult, if not impossible, to pinpoint exactly the fundamental difference between us and other species. Perhaps we should accept that there is no specific, clearly identifiable super-trait that is uniquely human. Perhaps we must conclude that our oversized brains simply accommodate a lot more of the same stuff that is present in animals. Perhaps we just happen to possess a great many rather ordinary characteristics, developed to an extreme degree. Or it may be merely the combination of all these capacities and characteristics that sets us apart, in which case, chasing the Holy Grail of proof of uniqueness is not all that interesting. Why should we need to prove something that stares us in the face? It is more interesting to find out what we ultimately are, how our minds function, and where the most salient brain functions are based—and, of course, how all of this came about.

There is no better starting point for such an enterprise than language, biologically the most spectacular defining characteristic of *Homo sapiens sapiens*, for there are no two ways about it: if a creature talks and understands, it is human. If it does not, it is not.[3] We are not concerned here with the specifics of any one language. Instead, we are interested in what we humans *do* with language and, perhaps even more importantly, what we would have been incapable of doing *without* language. The answers to these questions should give us some clues as to how long language has been with us. Next, what about the role language plays in our minds? How are language and thought connected, what are the interdependencies between these faculties and to what extent can they be considered one and the same thing? And last but not least, what is the nature of the link between language and consciousness?

These are big questions, perhaps too big. The answers cannot be proven: we do not have the witnesses, nor do we even have fossils to help us, for thoughts are immaterial, words ephemeral and our mushy brains and cartilaginous larynxes among the most perishable of bodily tissues. But there are certainly clues to be found, circumstantial evidence that will enable us to come up with a convincing case.

* * *

In a sea of uncertainty, there is one thing we can cling on to, and that is the fact that language is part of our biological makeup. It cannot be a purely learned skill, a cultural phenomenon, as behaviorists would have us believe. To be convinced of this, we need look no further than Chomsky himself, who drew attention early on to six points.

To begin with, there is the fact that children can learn *any* language. During the first few years of its life, every even remotely healthy child learns to speak the language used in its social environment, irrespective of where it is born or what particular language it is confronted with. We all admire and envy the ease with which child immigrants acquire the local language, unfazed as they are even by exposure to multiple languages at once. This goes for *all* children: in the many multilingual environments that exist around the world, children acquire the complete array of languages spoken around them as a matter of course. Two languages are common, but there are populations that make use of no less than five. And their children learn to understand and speak the lot, just as quickly as if there were only one.

From this, Chomsky concluded that what whatever was inherited, it was not the particular language or languages the parents spoke. If so, children would only be able to learn their parents' tongue, which is clearly not the case. Whenever a child is transplanted in a new environment, it automatically starts acquiring the language of that environment. Equally inconceivable is the idea that *all* languages are built in somehow. Alas, despite all their talents, children remain oblivious to the existence of the five or six thousand other languages spoken around the world today—not to mention the even larger number of languages that are now extinct. So Chomsky concluded that no particular language could be inborn. If anything was part of our biological heritage, it might at best be something enabling children to hook on to every possible language, as it were. A common core, a generic underlying scheme that could develop into any language depending on the input provided by the environment.

The second point is that *all* children embark on the language-learning path, and *all* travel along it at roughly the same speed. This strongly suggests that at least something is innate. Culturally transmitted skills, like playing the piano, knitting or dancing the tango, are qualities most children do not even begin to try to learn. And if they do, they seldom do so spontaneously, without an ambitious parent hovering in the wings. Moreover, few genuinely succeed. Would-be pianists and baseball stars all end up at very different levels of competence at vastly different speeds.

This ties in with the third observation. As long as children do not suffer from certain serious physical or mental handicaps, they *all* reach the highest level of competence: complete mastery of their language. This may sound strange at first. But the fact is that the huge differences in linguistic prowess between people are primarily social in nature. They have to do with how successful people are in putting their linguistic toolbox to use and finding the most effective, appropriate approach to the task or situation at hand. This, in turn, depends on differences in intelligence, concentration, self-confidence, education, training, experience and sociocultural background. However, the tools themselves are virtually identical for all of us.

Take someone who has trouble concentrating. Confronted with long and convoluted sentences, she will lose her way rather sooner than later. Remember what it was like at school when you were bored and tired, and you found yourself confronted with a dictation exercise? Further, people of lower intelligence may have more difficulty grasping the meaning and implications of some sentences, but that does not make them incapable of processing or producing a relative clause. Someone may have a hard time putting his thoughts into words, but rarely will he struggle to put his verbs and nouns in the right places.

Of course, the quality of the environment matters. Verbal prowess will inevitably suffer in children who are neglected or who grow up in surroundings in which linguistic ability carries little weight. Such children know considerably fewer words than the average child of the same age, and are also less aware of how language may be used to affect or persuade others. These are serious drawbacks, of course, but they concern people's aptitude to make socially effective use of language, rather than the actual mechanics of grammar and vocabulary. Once children reach a certain age, they can all understand any type of sentence. No single kind of subclause or grammatical construct is beyond them. As long as their memories are not overtaxed by overly complicated sentences, and they are not confused by a lack of knowledge of the world or particular words, they will manage just fine.

Sooner or later, all children start saying things they have never heard before. This was Chomsky's fourth observation. Yet strangely, these "originals" largely go unnoticed. We are much keener on those cute, precocious snippets of overheard adult talk children are wont to throw back at us: "Ah, isn't she a miracle? Just like Uncle Toby, and she's not yet three!" But these are no more than clever imitations. Children's less spectacular, truly self-made utterances are much more important, since they prove that language acquisition involves more than mere imitation.

Children get more from what they hear around them than is actually present in that input, so they must possess something that enables them to do so. In this respect, too, children do much worse when it comes to cultural skills. Whether it is playing tennis or the oboe, few pupils reach the level of competence of their instructors, and fewer still surpass it. Rare is the child who spontaneously acquires a particular knitting stitch, let alone invents a new one. Any child who pulls off a feat like that, in any cultural field, is rightly considered a prodigy.

The fifth point is that instruction by adults or other children is virtually useless. This is a good thing, too, because the language a child hears is language in action, full of mistakes, false starts, repetitions, and other shortcomings. Try reading a literal transcript of any real-world conversation. It appears that all a young child really needs to hear is a modicum of language. Deliberate training or instruction does little good; indeed, the pressure it generates can even get in the way of healthy, normal development. Cultural capers, on the other hand, depend on hours of dedicated training following explicit and elaborate programs of instruction, irrespective of talent. Enforcing a ruthless regime may not be every parent's cup of tea, but as the world of professional sports shows, it makes a huge difference.

Chomsky's sixth and last point, although by no means his least, was the sobering fact that the gift for effortless acquisition of new languages vanishes with childhood. Once puberty rears its first ugly spots, it is over and done with. This is when the plodding starts, when the gulf emerges between those with and without a flair for languages. This is when learning a language turns into the chore of memorizing words and sweating over conjugation tables, with all the frustration and failure that are part and parcel of the acquisition of cultural skills.

All this points not only to the innate, biological character of our language faculty but also to the existence of a *critical period*. This is a biological window of opportunity, a developmental phase during which brain and body are especially, even exclusively, sensitive to certain input from the environment. Not only language is subject to a critical period, and nor are such periods exclusive to humans. Take the development of sight in kittens. In a now famous experiment, newborn kittens that were kept in an environment with no vertical lines during their first weeks proved unable to deal with verticals for the rest of their lives. When they encountered a vertical line for the first time, the relevant parts of their brains had already reached their final level of development. Although there was nothing wrong with their eyes, they simply did not notice vertical lines, and were forever bumping into lampposts, door frames and table legs. A rather different kind of critical period is fertility in women, the interval from puberty to menopause during which they are reproductively responsive to male sperm. This shows that critical periods do not necessarily start at birth, do not have to end by adulthood and do not even have to involve the brain or mind.

As for our language faculty, there seems to be a large set of consecutive, partly overlapping periods during which children are sensitive to developing specific parts of the language system. Kids usually start producing their first words around their first birthday. But do not expect to hear them use pronouns like *I* and *you* until more than a year later. Things like the past tense are acquired at around three or four, while passives take another year or two after that, sometimes even longer. Not much can be done about this; trying to speed things up is fruitless. If a child has fallen seriously behind, some form of speech therapy or remedial teaching may help in catching up, but there is really no sense in trying to coax your child ahead of the natural development curve.

Taken together, these phenomena demonstrate beyond reasonable doubt that language is a real part of our biology. Think of it as a kind of abstract limb, just as much a part of an intact, healthy body as vision, hand-eye coordination, a sense of balance and a strong pair of legs to stand on. From the moment sperm and egg fuse, language is in some sense part and parcel of our genetic endowment.

This is not to say that everything related to language must be biological in nature. For one thing, we can keep on learning new languages long after the critical period has passed. We cannot do so as perfectly as in childhood, to be sure, and with equally little guarantee of success as any other cultural skill, but certainly well enough. There are even important aspects of language that lack any biological basis at all. Most conspicuous are reading and writing, but let us not forget good linguistic "manners": the vast array of subtle rules and precepts that dictate how to put things not only grammatically correctly but also in ways that are socially effective and acceptable. These are the rules that determine what counts as beautiful, literary and polite, or as rude, unsophisticated and disrespectful, and they differ considerably from culture to culture. In essence, language consists of a biological framework decked out with a bunch of cultural frills. Just the kind of mixed bag we ourselves are.

* * *

With few exceptions, biological traits are part of our genetic inheritance, stored in the DNA of the chromosomes we carry in each cell in our bodies. Which traits are coded in that DNA depends on one natural phenomenon and one only: Darwinian selection.

Ever since the very beginning of life, organisms have been created, shaped and reshaped, and destroyed again by the pressures of this process alone. Whether on land, under the sea, or in the very depths of the earth, the story is always the same: on average, those individuals of a species that happen to be best attuned to the circumstances of their day have more offspring. Deformed specimens, providing they even make it through birth and childhood, have few opportunities to procreate. Misfits and weaklings are prone to starvation, disease, accidents, and attacks from predators. Even if they survive all this, they get pushed aside in the race for meat and mates by their more robust brethren. The healthiest, strongest and most agile individuals always elbow themselves to the front of the queue. They are always on Club Nature's guest list.

This is why even mildly disadvantageous traits have always been mercilessly weeded out, and why, over generations, favorable characteristics tend to spread across the entire population in which they emerge. That no final, perfect equilibrium ever ensues is due to the eternal inconstancy of the environment in which species live. Climates heat up or cool down, become wetter or drier. New competitors enter the stage, or dangerous new predators or germs appear. Such changes in the scenery may render the most valuable of characteristics useless or even turn them into serious handicaps.

Selection comes in two forms. Natural selection is all about resilience in the face of prevalent natural conditions. The other form, sexual selection, is different but equally important to the continued existence of all species that use sex as their means of reproduction. This is about the qualities that make males and females attractive to each other, which in turn determines each individual's likelihood of scoring a mate. You might view the whole sexual rat race as a special case of natural selection, with the other sex as the single, all-powerful natural circumstance: be popular with the girls or guys, and Bob's your evolutionary uncle.

There is no reason why language should differ in some unfathomable way from all other biological characteristics. It, too, must have arisen through natural or sexual selection—or a combination of both. The big question, of course, is how this happened, because as evolutionary traits go, language is a very odd customer.

* * *

Although much of the language system is still shrouded in darkness, it is clear that we are dealing with a complicated system of highly developed, widely divergent possibilities and skills. Even disregarding the extensive, multifaceted cultural and social fringe, it seems far too articulate and convoluted to have come into existence in one fell swoop. Just think for a moment about how many things are involved in conducting an ordinary conversation. We easily produce more than 150 words per minute, each of them part of a highly structured pattern called a sentence. These patterns and the words that populate them must all be retrieved from memory, prepared and uttered at lightning speed, whereupon the hearer must re-engineer the whole thing into images and thoughts at the same rapid pace. Listeners are often so quick at processing that they outstrip the speaker—think about the last time you interrupted someone with an answer before they had even reached the end of their sentence. Miraculously, all this happens without us noticing any of the hard work that is being done. Even managing the scores of muscles in our face, mouth, and throat that enable us to produce well over seven hundred separate sounds per minute involves huge amounts of brainpower. Speaking is by far the most complex of all human motor skills.

And that is far from all. We have not even mentioned yet the apparent ease with which we fill in the gaps in what we hear, like when part of what someone says to us gets drowned out by ambient noise. We have not mentioned the Herculean task of building, maintaining and searching our mental lexicon. These personal dictionaries easily average thirty to forty thousand words and concepts in all their different guises, and in some people contain tens of thousands more. Despite their size, we retrieve any given word in milliseconds, including ones we have not heard or used in years. But we only ever notice our mental lexicon on the rare occasions when it fails to find an item for us in time. Lastly, we have not mentioned the very many snippets of socially acceptable code that help us approach others in just the right way, and which we all possess and learn to deploy with great accuracy, again, usually without paying any conscious attention at all.

Developing this beautiful, finely tuned apparatus must surely have placed a heavy burden on countless generations of hominids. Year after year, generation after generation, century after century, millennium after millennium, they must have spent vast amounts of energy maintaining those parts of our future language faculty that were already in place, fleshing out those that were budding, and anticipating the emergence of those that had not yet emerged. It must have required an immeasurably large and lengthy investment before this patchwork of skills and possibilities was complete and connected, ready to begin functioning as a coherent language-processing machine.

Practical considerations aside, none of this would be a serious impediment if evolution were orchestrated, and if it functioned like a modern, well-funded company with a competent staff of specialists and a clear idea of where it was heading. The idea for the "language machine" initiative would first be floated, then formally proposed and carefully discussed, planned and modified at the highest level. Once approved by the board of directors, considerable time and man-hours would be allotted to its development, with a budget to match. Jubilant press releases would be issued, heralding the forthcoming Language Machine® by Darwin's Evolutionary Aggregates Ltd (DEA). Ministers, heads of state and other hotshots would be given a tour of the laboratories and testing facilities. The task would be far greater and more costly than putting the first man on the moon, but given enough time and funding, the boys at DEA just might pull it off.

This, of course, is not how evolution works. There is no company, no board of directors and no capital for long-term investment. Nor is there the slightest trace of planning. Instead, Mother Nature just monkeys around. And a good thing, too, for in a world like ours, in which circumstances change unpredictably, long-term planning is a recipe for disaster. It is much better to experiment as much as possible, with sheer chance as your guide, producing bucket loads of failures and the occasional gem. This is precisely what nature does: the larger the variety of experimental versions of an organism, the better the chances that one or two will be sufficiently well attuned to the accidental circumstances into which they are born, to ensure the survival of the species. Every small step, every tiny change in the make-up of an organism is immediately

Charles Darwin (1809–1882), father of the theory of evolution. This picture was taken in 1855, four years prior to the publication of *On the Origin of Species*.

put to the test in each and every individual who is affected by it. There is no such thing as proof of principle. There are no mock-ups, no developmental laboratory models, no testing versions or prototypes. Evolution is a sidewalk engineer, an opportunistic tinkerer.

Succinctly put, Darwin's theory is about the "survival of the fittest." This phrase was coined in 1851 by the British scholar and liberal philosopher Herbert Spencer in his book *Social Statistics.* Spencer would play an important role in disseminating the ideas of *The Origin of Species* after its publication in 1859, not least because Darwin himself adopted Spencer's apt adage. But as striking as it is, it is also quite misleading. Survival of the fittest suggests that nature works like a sports tournament, a ruthless competition in which glory goes only to the very, very best, leaving nothing but failure for the overwhelming majority. Fortunately, this is not the case. In order to survive, you do not need to be among the extremely well-adapted elite. Evolution is much more like school: to move up a grade, you only need to pass the exam, and a "B-" will do just as well as an "A." In evolutionary terms, you have got that "B-" in the bag as long as you are just fast enough to outrun powerful enemies, just strong enough to ward off lesser ones, just healthy enough not to succumb to bad weather and typical ailments, and internally just efficient enough to make do with the resources left to you by competitors.

What this means is that every change must yield at least enough of an advantage in the struggle against others and the elements to balance its energy costs. It is clear that mutations that harm their bearers early in life, like those causing hereditary diseases or malformation, have little chance of spreading among a population. The afflicted individuals will either perish before they reach sexual maturity or be disadvantaged in the race for mates. But a mutation that *seems* advantageous will also set its bearer on a course toward an untimely demise if it demands too much energy, requiring him to eat more than his body can cope with.

Mutations that do not seem to have much effect either way stand a good chance of entrenching themselves snugly in a population and settling in for the long haul. But they may also just as easily vanish into thin air again. These apparently ineffectual variations in the structure, appearance, and behavior of an organism are called genetic drift. When circumstances change, such variations may suddenly become highly advantageous—or seriously detrimental. Relevant changes in the external environment may be prolonged drought, say, or the emergence of a dangerous virus. But the organism itself may also change. For instance, it is quite possible for two innocuous, existing mutations to combine with a new one in a way that affects the efficiency of the organism.

When all is said and done, genetic drift is a kind of deferred selection. It is a process by which small, virtually meaningless changes can accumulate in an organism until at some point they suddenly start having real consequences. It is a bit like mixing together sand, salt, and

Herbert Spencer (1820–1903), a prominent British liberal philosopher who coined the phrases "survival of the fittest" and "social Darwinism."

sodium carbonate. Under normal circumstances, the result is nothing but a drab mixture of mineral grains. But add a great quantity of heat and suddenly it transforms, as if by miracle, into something spectacularly different: sparkling, transparent glass.

Likewise, ever so slowly and innocently, genetic drift can lead to spectacular changes. But the mutations involved must be all but neutral in terms of the energy they require. This makes it highly unlikely that genetic drift is behind the emergence of the human language faculty. Our brain is an inveterate glutton; even at rest, it gobbles up one-fifth of our total energy expenditure. So building additional brains and kitting them out with more and more structure comes at a high cost. This cost is justifiable if it brings immediate advantages, but what would a hominid gain from a quarter of a language faculty? Or half of it? Although genetic drift quite likely played some part in the genesis of human language, it cannot account for all of it, not by a long stretch.

There is another, even more serious problem. It is usually taken for granted that the evolutionary advantage that justified the development of human language was its communicative function. This may seem only natural. After all, what is language for? It is to blab with others about anything and everything, of course, up to and including our deepest sentiments. When we look a bit closer, this assumption turns out to have consequences so bizarre that it simply cannot be right.

Human language does not consist of a limited range of biologically built-in, immutable messages, as is the case with animal communication systems. The very essence of human language is its capacity to allow for infinitely many different messages, by combining the tens of thousands of words we have at our disposal in ever-changing ways according to a limited number of rules. What makes this possible is precisely the *absence* of the direct, inseparable link between message and meaning to which animals are bound. In human language, a thought or feeling is first encoded in words in line with the rules of some language, and then transmitted, whereupon the receiver must decode the message using the same rules and her own, sufficiently comparable lexicon. So communication through language depends on having a shared system of rules and a roughly matching vocabulary. To communicate, people must literally speak the same language. Now if language were developed for the purpose of communication, our ancestors would only have profited from it once others around them had an operational language faculty too. But this, evolutionarily speaking, is a bridge too far. A trait that only improves your chances if *someone else* has it too just does not make sense when it comes to natural selection.

* * *

And so the mystery deepens. If communication is not the driving force behind human language, then what is? What could have justified the development of the intricate, highly interconnected skills and possibilities that result in language? What evolutionary advantage does it bring for each of its individual owners, independent of others? And how long ago did this come about? Before we can really start digging for answers to questions like these, we must be very clear on exactly what we mean by the term *communication*. And for this, we need to look at how human language relates to "animal languages": the range of expressions animals use to make themselves understood to others—expressions that clearly did evolve for their communicative advantages.

Notes

1. Chomsky clung to the existence of this mental language organ far longer and more tightly than seemed reasonable. The louder the protests, the more fervently he appeared to insist on its unique and dedicated character. This may be attributable in part to an aversion to anything smacking even remotely of behaviorism—after all, Chomsky made his name in 1958 with a devastating critique of a book called *Verbal Behavior* by the then-god of behaviorism, B.F. Skinner. However, taking into account what Chomsky was to say much later, I suspect that to some extent it was also simply a hapless case of mutual misunderstanding. Chomsky routinely refused to discuss the subject of the reality of the organ, for lack of debatable arguments. His critics interpreted this as ducking the issue, to which cocksure Chomsky, used to being the brightest kid on the block, responded by digging in even deeper.

   It is revealing that many critics riled him for calling the origins of language a mystery, a position they considered unfit for a true scholar. What they failed to appreciate was that the Chomskyan idiom distinguishes between *questions* and *mysteries*. For Chomsky, *questions* are the kinds of problems that scholars can use their toolkits to tackle. *Mysteries*, in contrast, are matters that nobody has even the slightest inkling how to deal with, and are thus best left alone until further notice. For many years Chomsky refused to consider the origins of language for the very same reasons that NASA has not invested seriously in expeditions to the next solar system: nobody has the vaguest notion of how to go about taking even the first steps. It was in this sense that Chomsky held that the origin of language was a mystery.

2. This is not to say that we cannot use proper names as signature whistles. Shopkeepers, doctors, lawyers, and companies do call out their own names all day long by means of shingles, logos and advertisements. But even in these cases, the idea is not primarily to identify oneself, but to give the observer a sense of recognition and familiarity. Think of the yellow arches that identify McDonald's all over the world, the red and yellow Shell emblem or the Union Jack that British soccer fans paint on their faces.
3. Technically, not even this is absolutely true. Deaf people who use sign language do not actually talk, and certainly do not listen by means of the "usual" channel. Yet they are just as human as the rest of us. Certain parrots "talk," as do some mynah birds (which are related not to parrots but to sparrows). And some species of animals, like dogs, even "listen" to commands. Yet none of these animals has language, nor are they human.

2

The Animal Inside Us

What to do if you are young, ambitious and a biologist, and you want to discover something really new? You make for New Guinea. On this vast and inaccessible tropical island, unexpected danger still lurks around every corner, even in seemingly innocent creatures, as University of Chicago graduate student Jack Dumbacher discovered in 1989 while researching the lives of birds of paradise, high up in the New Guinea mountains. By accident, a quite different bird had become entangled in his trap net, around the size of a sturdy blackbird and just as black, but with a bright red breast, back and tail. Gently, Dumbacher set about disentangling his unwanted catch, but the frightened animal kept pecking and scratching him. So intensely did it fight that at some point Dumbacher, stifling a curse, stuck a painfully damaged finger in his mouth to stop it from bleeding.

To the growing alarm of our wounded bird catcher, within seconds his tongue and lips began to tingle. There was a numb, burning sensation in his finger, which also spread across his tongue and mouth. Only after a few anxious hours did the sensations begin to subside. Later, it turned out that Dumbacher had stumbled onto a remarkable discovery: the mountain pitohui—that was the name of the bird, on account of its pee-twee-like chirping—was poisonous. It was well known that many species of insect and fish were venomous, and so too were lots of snakes and frogs. But a bird armed with poison in its skin and feathers was unheard of. Curiously, the poison proved identical to that which makes South American poison dart frogs (*Dendrobatidae*) such unpleasant company. And just like dart frogs, it seems that the pitohui do not produce the poison themselves, but ingest it with some form of prey. This may explain why the degree of toxicity of the birds varies considerably from place to place.

Dumbacher's assailant represented just one of no fewer than seven kinds of pitohui. Three are now known to be seriously venomous, and of these two have quite conspicuous plumage. In this respect, these

The poisonous pitohui of New Guinea.

unusual pitohui do something that is very common in the animal kingdom: they advertise their unsavory properties. In many animal species, bright, gaudy colors send a clear message: "Beware, I am dangerous: I'm poisonous or otherwise unfit for consumption." It is this strategy that makes us wary of wasps, garishly colored caterpillars and toadstools in glowing, unearthly hues. Even some birds use this kind of signaling, to the advantage of all parties concerned—besides the unlucky Dumbacher, of course. Conspicuous looks would seem to be nature's most widespread form of communication: hundreds of animal and plant species use this technique, and even more get the message.

Yet, this is not the type of thing we usually have in mind when we talk about communication. We tend to think of communication in the sense of how *people* use language, and in terms of modern means of communication like radio, television, and the telephone, in short, how people and man-made devices communicate with one another.

Communication experts break this down into four elements. First, there is a sender. This sender encodes a message—element two—and sends it on its way. The message travels through a medium, the third element, and is ultimately picked up by element number four, a receiver, who decodes the message again. The properties of the code are such that messages can travel easily and swiftly through the medium of choice.

In the case of human language, the typical medium is air, and the sender encodes messages in the form of sound. Sound travels well through air. For radio and television, the medium is the ether, which is unfit for sound. So we use machinery to translate messages into electromagnetic radiation and back again. The classical telephone works by means of electrical pulses that travel through copper wire, or light pulses racing up and down glass fiber cables. Old-fashioned semaphore systems, dating back to Napoleon's day, encoded messages into patterns of light, which could be seen and interpreted from a distant hilltop or steeple.

This take on communication contains at least four presuppositions that are irrelevant to what the pitohui, the wasp, and all those other little communicators do: the assumption of some kind of intention, sharing knowledge, coding and reciprocity.

First, the sender-receiver model implies that the sender has a premeditated purpose. He wants to get something across, and so encodes information and sends it on its way to a recipient. These are usually conscious activities. Or at the very least, the message that is sent forms a specific reaction to some prompt. Such prompts need not always be obvious from outside. The urge to communicate something may just as well originate inside the sender's body. Think of hunger, or pain, relief or elation. The point here is that we rarely send messages for no reason or without being aware of it. The poor pitohui, by contrast, is born with its garish plumage. It neither intends to send its message nor purposefully encodes it. The bird transmits its warning message in spite of itself, just as do wasps, caterpillars, and toadstools—or, for that matter, traffic signs.

An equally natural presupposition of the sender-receiver model is that communication is a means for sharing knowledge. The sender knows something the receiver presumably does not—and "knowing" may also apply to feelings here, including sensations like fear, hunger or pain. The sender then sends that something in the form of a message to the receiver, who from that moment on shares the encoded knowledge with the sender. Therefore, communication makes sender and receiver more alike, at least as far as their knowledge is concerned. This, too, differs from the pitohui and creatures like them. Members of other species may appreciate the cautionary value of a conspicuous exterior, but the pitohui themselves lack even the faintest notion of their own unsavory properties. What is more, since very few species indulge in cannibalism, there is little need for these animals to even understand their own message. To themselves and their conspecifics, it is meaningless beyond, "Hey, another pitohui!"

Finally, there is not the slightest hint of reciprocity, of signaling something in return in order to acknowledge receipt of the message or maintain a channel of communication. Wasps and caterpillars, toadstools and others that wear warning colors share nothing with those who perceive them, let alone a common code that would enable them to engage in any manner of back and forth. In fact, they do not even want contact; the very point is to avoid it. Striking as they may be, the true meaning of the colors of the pitohui, the stripes of the wasp and the luminous hues of toadstools is nothing more than "get lost, nothing to see here."

* * *

All told, it is a measly affair, the communication between unwholesome prey and their potential assailants. In fact, so is virtually all signaling by animals and plants to other species. In the great majority of cases, such communication serves not to promote, but exclusively to avoid closer contact. It makes good evolutionary sense not to get in each other's hair unnecessarily. Sometimes signals are used purposefully to scare others off. Think of all the howling and growling, the shrieking and hissing with which animals welcome approaching strangers. But these cases are the exception. Much more often, there is no intent. Most communication to ward off contact is of the passive, involuntary kind of the pitohui and the wasp.

When animal communication between different species does serve to make contact, it is usually the kind of encounter that leads to one party being eaten by the other. The sweet scent of putrefaction that carnivorous plants use to lure insects is a prime example. To flying small fry, it is as seductive as the sirens were to Ulysses, and just as lethal, or more so, since the great Greek hero narrowly escaped, whereas the poor insects are almost always doomed. There is just one exception to this rule, and that is the use by trees and plants of colorful, fragrant flowers and tantalizing fruits to attract insects and fructivores, which then do the work of pollinating and disseminating seeds for them. This is the mother of all win-win situations: the animals are nourished by the plant, and unwittingly return the favor by ensuring the continued existence of countless species of herbal life.

Communication between members of the same species is barely richer than what goes on between species. Much of what animals signal to members of their own species is, likewise, meant to keep contact to a minimum. When two cats in a backyard are hissing at each other, fur on end and tails erect, they are sending but a single message: "Clear off, or face the consequences." Although centuries of relentless breeding have seriously watered down their feral genes, dogs still spend most of their time outdoors peeing against trees, poles, and whatever other landmarks they come across. This is undiluted communication, staking out a territory by means of olfactory markers. "This is my patch, I'm the boss around here," is the message, and competitors are well advised to take notice. Coming from our pampered house dogs, this may amount to little more than tokenism, but among undomesticated animals these things are taken very seriously. In the wild, if you do not respect territorial boundaries, you flag yourself as an unfriendly trespasser, and you can count on being treated as such.

Expressing threats and warnings is probably the most frequent use to which animals put their communicative powers, even to members of their own species. This is not to say that animals generally treat each other poorly. Animals that live in groups also warn one another of danger, sometimes even putting themselves at risk by doing so. Meerkat communities are a good example. These small predators of the Kalahari Desert always have a number of sentries on the lookout. They stand stock-still, straight as a broom, monitoring everything that moves. When they spot something untoward, they immediately alert the whole troop with a whistle ("Attention! Possible danger approaching!") or a bark ("Run for your life!"). By doing so, they betray their own position and put themselves in harm's way—but this is just part and parcel of the life of a meerkat.

Many fish species, too, warn their fellow fish when they come under attack. If their skin is torn—say, by the teeth of a predator—alarm pheromones are released into the water. This is a type of scent molecule, and when nearby fish get a whiff, they know exactly what it means: "I'm done for—save yourselves!"

Even certain plants can warn their neighbors when they are being invaded by a parasite. They send out chemicals that waft over to their untouched peers in the vicinity, which, in turn, take this as a cue to start arming themselves with repellents and poisons.

Diverse as they are, these means of giving fellow creatures a chance to save their own skin are much closer to communication in the usual sense than the shining armor trick. There is a kind of intent, in that they consist of specific reactions to specific impulses. Also, information is conveyed: sender and receivers come to share awareness of a danger that only the individual under attack knew about before. But nothing is coded. The signals are bound to one and the same meaning, and the receivers always react immediately and in exactly the same way. No dialogue ever ensues to check facts or go into detail: "Three bad guys, you say? By the fern? The evergreen or the one with the purple flowers?" There is no reciprocity, no response, not even so much as a "by golly!" or a "thanks for the heads up!"

* * *

Animal communication may, of course, have other uses than deterrence and warning, notably within small, closely knit social networks of families and troops. Animal families are the bread and butter of nature films; producers love them, because the audience loves them even more. This is mainly because animal families display behaviors that are extremely rare elsewhere in the animal kingdom: kind-heartedness, tolerance, care and consideration, and playfulness. These are precisely the kinds of things that we like to encounter in our fellow man, and perhaps see too little of. It would seem that great numbers of people fill the void where idyllic family life ought to have been by watching father lion good-humoredly let his cubs climb all over him, doling out a lazy-pawed slap when their tomfoolery gets out of hand. Or they peep through a webcam at a pair of songbirds gallantly trying to quell the insatiable hunger of a nest full of tweeting fluff balls. We also see lots of communication going on in such families, some of it real, some only because we wish to see it. We even see education and training in action. We see lion parents "teaching" their cubs to hunt, and young birds receiving "flying instruction," and we readily imagine the accompanying commands and parental admonishments. We see animals behaving almost like human beings, and that strikes a chord. But for the most part, we are deceiving ourselves. This so-called education, however indispensable it is in terms of development, rarely amounts to anything more than giving the young a chance to put into practice their instinctive repertoire, not unlike the way human cubs learn to walk and use their mother tongue with virtually no explicit instruction.

At best, animal education involves learning by imitation, but even that hardly counts as full-fledged communication in the sense of the sender-receiver model. Learning by imitation does not involve coding; rather, it is straightforward copycat behavior. What is learned is identical to what is observed. Nor is there any reciprocity. Animal education never develops into a game of questions and answers, of trial and correction. This kind of knowledge transfer is again strictly a one-way street, from teacher to pupil, and the pupil never has to sit an exam or endure a progress evaluation.

Animal families are temporary constellations, because sooner or later, children grow up. For solitary species, reaching adulthood also puts paid to most social life, and with it to virtually all forms of communication except threats and warnings. Things are different for animals living in herds, packs, swarms, tomes, pods, prides, coveys, bevies, bands or whatever else their groups may be called. Yet even their communicative repertoire scarcely exceeds the bare necessities.

The diminutive bee can be considered one of the better communicators. Its "waggle dance" is probably the most complex symbolic system so far deciphered in the animal world. When they have discovered a new location to colonize or a new source of flowers to pump for their nectar, water and waxy building material, bees dance the waggle to reveal to the rest of the hive the whereabouts and quality of their find. The scout bee, as she is known—and it is always a she, as drones spend their days loitering about, waiting for mating season—dances in a figure of eight on a vertical honeycomb. The other bees can deduce the direction toward the discovery with respect to the sun from the angle of the dance with the vertical, which represents the direction of the sun as seen from the entrance to the hive. The speed of the dance indicates the flying distance, and the vigor with which the scout wiggles her rear end illustrates the quality of the find.

This comes close to fitting the sender-receiver model. There is intent: bees only dance when they actually have something to report. Also, knowledge is being shared. By observing the dance, the other bees learn something that only the scout knew, and within the narrow limitations of the bee repertoire that knowledge is always novel and unpredictable. Reciprocity is lacking, in that the other bees never dance something in return, never ask for further directions or explanations. But there is certainly a coded message. The bee dance, considered by many as one of the seven wonders of the world of animal behavior, is made up of symbols: gravity symbolizes the direction of the sun, speed stands for distance, and vigor represents the richness of the find. Beyond that, however, the meaning of the bee dance is tightly constrained. It is *always* about newly discovered sources of food, water, or building materials or potential nesting locations, and it only ever indicates direction, distance, and richness. No scout can ever go into more detail than this.

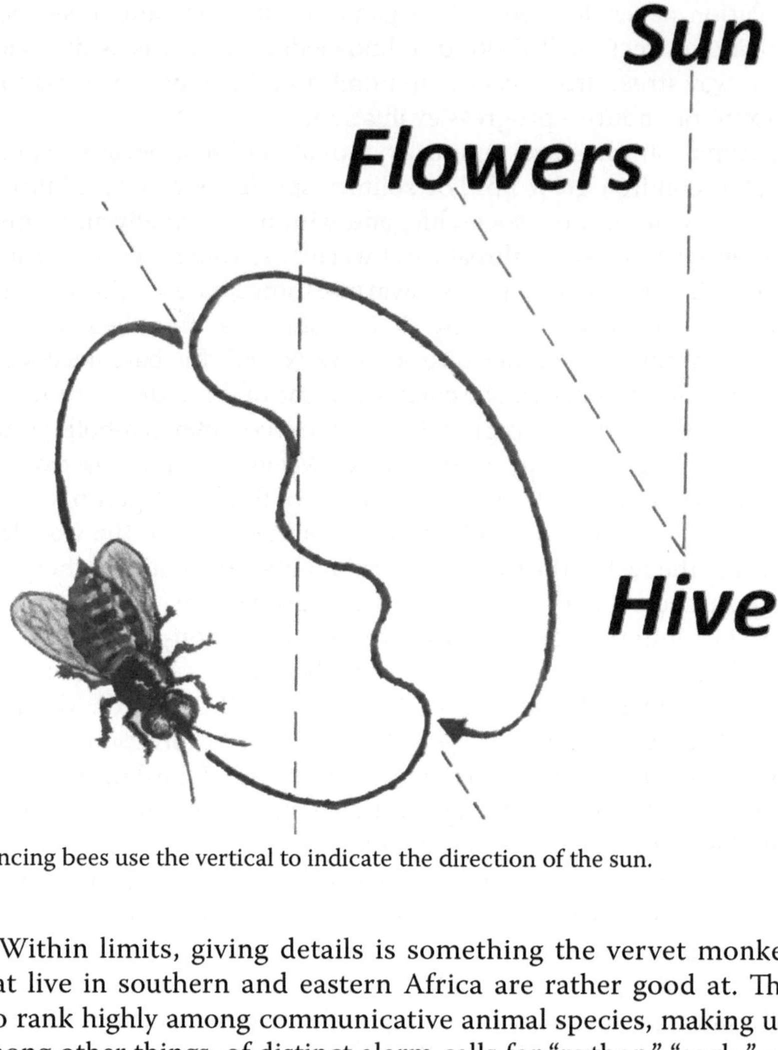

Dancing bees use the vertical to indicate the direction of the sun.

Within limits, giving details is something the vervet monkeys that live in southern and eastern Africa are rather good at. They too rank highly among communicative animal species, making use, among other things, of distinct alarm calls for "python," "eagle," and "leopard"—all predators that do not mind a bit of vervet for lunch. Interestingly, these calls must in part be learned. Even very young vervets can produce them perfectly. That much seems to be innate. But youngsters use them in somewhat more general ways, as if they stand for "danger in the grass," "danger from the air" or "danger on the ground." As they grow up, they learn to use the python cry only when it really is a python lurking in the grass, the leopard bark only for an actual leopard or similar feline, and the air-raid alarm exclusively for eagles.

Vervets have yet another rare specialty: they fool one another. Sometimes they will shout "python," "eagle" or "leopard" for no apparent reason other than to scare all their nearby peers into frenzied flight. This is an ideal means of minimizing the risk of prying eyes when they have found a delicious morsel of food they are not inclined to share. For us, it proves that spreading false information on purpose is by no means an exclusively human activity, as was long thought. Beyond this, though, the communicative powers of vervet monkeys are hardly impressive. Their few cries are each rigidly bound to a single, immutable meaning and, save for a little refinement, completely wired in from birth.

Deceit in vervets is fairly innocuous, something that cannot be said of that master swindler among animals, the cuckoo. Quite shamelessly, the cuckoo lays its artfully disguised eggs in the nests of other, smaller bird species, saddling them with the task of hatching and rearing its young. And it stops at nothing to ensure it reaps all the benefits of this intrusion. Cuckoo eggs typically hatch a few days before the eggs of their involuntary foster parents. Or rather: before they would have hatched, because the hatchling sponger rolls the rest of the eggs out of the nest before they do. That way the parents do not get suspicious: with no legitimate offspring around for comparison, they cannot see that their chick looks a tad funny. For good measure, cuckoo young also sound exactly like the true offspring would.

That is the idea, anyway. Different kinds of cuckoos do not just drop their eggs in any old nest. Rather, each favors its own particular species of foster parent, and hatchling cuckoos can produce the typical calls of these foster parents flawlessly and from the get-go. Clearly, the calls are innate. Sometimes, however, a cuckoo will lay its egg in the nest of a less preferred species. In this case, the hatchling needs a few days to perfect its imitation of the foster parents' chicks' call. This is an extraordinary achievement—after all, it has personally seen off its stepbrothers- and stepsisters-to-be, so there is no one left to eavesdrop on to get the sound just right. So how does the young cuckoo pull it off?

The answer was not discovered until 2008, by a group of Australian ornithologists. After hatching, the young cuckoo soon notices that its cries are alarmingly ineffective—its unwitting foster parents just do not react with the desired enthusiasm. This is when it begins to throw some variation into its built-in repertoire. As soon as some of these calls turn out to have the right effect, it goes on to hone its performance until, in a mere matter of days, it reaches a perfect imitation of the true offspring of its foster parents. The hatchling has every reason to work hard at this, for the reward is sweet and essential: the better it does, the harder its foster parents will work to fill its hungry belly. The process is not vastly different to what people the world over do—except native speakers of English, of course—when confronted with foreigners who speak an unfamiliar tongue: "*Doitsch? Fransei? Espaniol?*" Only when the other party starts nodding happily in recognition do they begin conversing with them in that language as best they can. And when they do, they get better at it, albeit much more slowly than the cuckoo.

This is not the only difference between us and a cuckoo, of course. Whereas we are wont to start conversing about all and sundry, a young cuckoo invariably harps on one point only: it wants food and more food. For all the noise it makes, it is not so much communicating as performing an act of mimicry. Its repertoire is set in stone, immutably anchored in its genes. The best it can do, if the circumstances demand it, is to vary this slightly.

* * *

More sophisticated cheating and deception can be found among our closest cousins, the great apes. They have an extensive communicative repertoire. Grooming, grimacing, making noises, posturing and, among bonobos, even mating and mutual masturbation all serve to maintain good relations between group members and keep order in the social pyramid. And there is more—much more than any other animal can boast of, as far as we know. Chimpanzees forge coalitions and hatch plots, as the Dutch ethologist Frans de Waal first reported in the early 1980s. Then, having accomplished what they set out to do, rivals expressly make up again. Also, some individuals are consistently protected, whereas others do not seem to be able to do anything right. These are all indications that ape societies are complex enough that considerable intelligence is required to survive in them.

As animals go, apes do possess exceptional wits—this much is clear from the ways in which they deceive one another. There is a well-known story about a bonobo that had been wounded in a contest with the incumbent alpha male. For months after it had recovered, it would take to limping ostentatiously whenever it was within the range of vision of the unbeatable bully. Or consider the case that always appeals most to us humans: the stealthy-shag-behind-the-boulder trick. Out of sight of her lord and master, a female is busy being mounted by an attractive but less prominent male. He, too, takes care to ensure that all compromising body parts are out of the boss's line of sight, while grinning innocently at him from behind the rock: "Everything's hunky dory over here." It is clear that such interlopers understand what the big man can and cannot see from his point of view, as did the limping bonobo. Whatever our fascination with such apish adultery reveals about ourselves, their pulling the wool over each other's eyes certainly betrays something important about chimps and bonobos: they have an idea of what the world looks like to *someone else*. This is known as a theory of mind, and it is extremely rare among animals.

Apes also indulge in explicit educational activities. They teach one another how to handle tools, for instance, like sticks for getting at hard-to-reach insects and honey. They even instruct one another, to some extent at least. There is an endearing film clip dating from the 1930s of two young chimps joining forces to pull a case of bananas toward them. Coordinated cooperation is of the essence, since the case is far out of reach and all that can help is a loose rope that runs from their cage around the case and back again. They can reach both ends of the rope, but if only one of them pulls, all they end up with is the rope—close maybe, but no bananas. The only option is for them to agree on each taking hold of one end of the rope and pulling together.

This is not an easy task, for it is far beyond anything the animals encounter in nature. Their natural habitat simply does not throw up cases of bananas with ropes conveniently lashed around them. Contemplation and analysis are required, as is communication and consensus on the plan, followed by coordinated effort. And they do it: pulling neatly in time, like sailors hoisting sails with a smooth, rhythmic cadence, the two young chimps bring their sweet reward closer and closer. What is more, whenever one of them loses interest and starts to wander off, the other buttonholes him by pulling his fur, forcing the would-be deserter to keep pulling for a little while longer. This is beyond doubt the closest thing to human behavior you will ever see in apes, but as so often, a caveat is in order. It is unclear from the film to what extent the apes found out for themselves what to do, and to what extent they were instructed. We can just make out that they are both given one end of the rope, and then a caption follows, which reads, "With further practice the performance becomes smooth, rapid and uniform." This suggests considerable instruction and training by human caretakers.

Nevertheless, the fact that they can learn how to perform a task like this at all shows that the great apes are much closer to us than traditionally thought. We can even find traces of culture, snippets of noninnate knowledge that are transferred between individuals, from generation to generation. Chimp researcher Jane Goodall was the first to point this out in the early 1970s, after observing the intense efforts of the young chimpanzees in the Gombe reservation to learn to handle tools by copying their elders. Soon after, reports emerged on regional differences in grooming techniques, social behavior, and ways of gathering food among wild chimpanzee troops, as well as variations in tools and tool use. These were mostly attributed to genetic variation: not culturally induced, but innate behavioral differences that seemed to be linked to genetic variation between troops. Recently, however, statistical wizardry has given support to the idea that the genetic variations do not tally with the behavioral differences that have been observed. If this is correct, then ape societies really do show evidence of rather subtle forms of cultural knowledge transfer, long considered an exclusively human phenomenon.

* * *

Yet, however intelligent and blessed with communicative powers they may be, apes show no trace of anything remotely like human language. Theirs is no more than a rich variant of the kind of communicative system that we encounter in thousands of other species. And all these systems are characterized by the same design properties.

First, every expression is immutably linked to its meaning. If an animal expresses submission by way of a particular grimace, then all individuals of the species express submission in the same way, and the grimace means the same thing all the time. There is no variation among individuals, and nothing changes between generations. Coding is virtually nonexistent, and for good reason: if there is a cast-iron link between expressions and meanings, then expression and meaning are one and the same thing.

In many cases, it makes little sense to even talk about meaning. Take the pheromones flowing from the injured skin of certain types of fish. Other fish do not have to recognize that they signal imminent danger, let alone deduce that there is a dangerous predator around. It is quite enough for them to smell a foul odor, just as we experience the stench of a turd, to elicit the effective response: yuck, back away! Swimming away from the pheromones means putting distance between oneself and danger, regardless. By the same token, it is immaterial whether an animal simply reacts mechanically to a warning cry or actively realizes, "Uh oh, predator approaching!" As long as it makes off or hides, it maximizes its chances of survival, and from an evolutionary perspective this is enough.

Vervets seem to show a little more sophistication in this respect, with their specific calls for "python!," "leopard!," and "eagle!," but not much. To react adequately, other vervets do not need to understand the finer shades of meaning. It suffices to take these calls to mean, "Run, and watch out for movement in the grass," "Run, and climb a tree as fast as you can," or "Run, and make for the shrubs," respectively—just as young vervets understand them. Yet, as they age, they do attribute more specific meanings to these three cries, so some real coding does seem to occur. Why this is the case remains a complete mystery.

Some animal communication systems can boast of combinations of signals. But these, too, are set combinations with a single, integrated meaning. They do not permit variation, and all constituent parts are expressed simultaneously. For example, a dog threatening with attack lowers its tail while at the same time raising the hair on its neck, curling its lips into a snarl and producing a low growl. No dog in a similar situation will ever growl this way *without* snarling or while wagging its tail enthusiastically. Nor will any dog ever express the combined signal to convey anything but a final warning to back off. The whole conglomerate of signals is an indivisible whole, an integrated expression of the animal's physical and mental state at that moment. It is not the sum of a number of independently meaningful units of expression, as is the case with human sentences. The signal and the disposition that gives rise to its expression are one and the same thing. There is no distance, and therefore there can never be anything like irony or hyperbole. Even understatement is absent, except in cases in which animals were expressly bred for it. We see this in certain fighting dogs, selected over many generations for their tendency to attack without prior warning. The very success of such breeding programs shows the high degree to which animal communication is genetically driven.

It is not easy to find exceptions to these rules of animal communication. Perhaps the ape stealthily fornicating behind a rock is one, for the innocent face he presents to the alpha male does not match the activity in which the rest of his body is involved. If so, any exceptions remain limited to the intellectual cream of the animal crop, our closest relatives.

Another characteristic of animal communication systems is that they are not bound to any single channel of expression. Grimaces, postures, sounds, scents, gestures, touching, and ritualistic performances, like the courtship dances found among birds or complex posturing aimed at intimidating rivals, may all form part of the system of a single species. And almost any mode of expression can be marshaled for communicative duty. Californian ground squirrels, for example, even use infrared radiation to communicate with rattlesnakes. These squirrels are fierce fighters, and immune to the snake's poison. If a snake gets too near a squirrel's young, its bushy tail heats up. Rattlesnakes have heat sensors in their noses—they smell heat, as it were—and understand the squirrel's signal as a clear warning: "Go ahead, punk, make my day!" Other species of snake, like the gopher or pine snake, do not have these heat sensors. So when a squirrel chances upon one of these snakes, its tail does not heat up—there would be no point.

Tail temperature is rather an exotic means for getting a message across. Much more common are variations on voice, tail movement, fur and posture, and of course urine and excrement. But regardless of the number of different modalities in an animal's communicative system, not one can fall by the wayside or be replaced by another. No dog or other animal can choose to substitute a urine marker for some other signal, such as a sound, a pose or some pungent glandular chemical. A dog whose tail has been docked will not even seek to compensate for its wagging. The link between signal and meaning in animals is immutable, forever and always.

Finally, animal communication systems offer little room for individual variation. Most animal communication is about threats, territorial rights, and the pecking order, and in such matters, clear and unambiguous communication is paramount. Individual frills can only lead to misunderstandings. What little variation there is usually lies in the specific sound of an animal's voice, which some species use to distinguish between individuals. Variation in the signals themselves is almost exclusive to aquatic mammals and to songbirds, which create their own, sometimes quite sophisticated individual repertoires. But even these songs do not express different messages from those sung by others. They serve merely to advertise the singer as the loveliest, healthiest, most alluring male around. So birdsong, in spite of all its variability and complexity, is no different from other animal communication systems. Each song has but one meaning, which is indivisible

Jaws wide open, ears back, and bristles out: clearly, no good is to be expected from this Bengal tiger.

and immutable: "Come and get it, girls! I'm Supertit! I'm Alphafinch! I'm the blackest blackbird of them all!"

* * *

Unlike animal systems, human language is very much bound to a single communication channel and one specific format. All our utterances are sentences, each a cleverly composed lacework of content words interwoven with words and particles that serve grammatical purposes. How these sentences actually come out depends on the channel and format we are using, that is, a particular language and a particular mode of expression, be it spoken language, sign language, or writing. But once we have chosen a language and a mode of expression, we stick to it, at least for the duration of a given communicative act. We do not normally switch languages mid-sentence, or say half the sentence out loud and write the rest down. There is no sharper contrast between this singular discipline of human language and the chaotic, unfettered free-for-all that characterizes animal communication, cavalierly throwing together a hotchpotch of expressions to form a single signal, from scents and grimaces to things as outlandish as heat radiation.

At the same time, animals are hopelessly tethered to their very limited mishmash of set messages and ways of expressing them. Human language, on the other hand, not only allows for the formulation of infinitely many different messages but can also make use of entirely different channels. The deaf express themselves just as well in sign language, using their hands and bodies to make messages and their eyes to perceive them, as hearing people do using their mouths and ears. And acquiring a sign language is no more difficult than learning any other foreign language. If you do so at an early age, because of, say, a deaf parent or sibling, you will pick it up as easily as any other language with which you happen to come into close contact. You simply become bilingual in English—or whatever language it is you speak—and sign language.

These days, children learn to read and write in many parts of the world. This is a great example of switching channels, even more so than sign language. A sign language not only uses a different channel but it is also a full-fledged language in its own right—a member of a different language group, in fact, given that sign languages differ from community to community. Reading and writing are not about using a separate language, but about translating a spoken language into a different channel.

Further, the sentences of human languages, sign languages included, are not monolithic blocks. Quite to the contrary: they are made of combinations of smaller units, each of which has its own meaning. The meaning of a sentence is the sum of the meanings of its parts and the order in which these parts are put together. In any given sentence, words can readily be substituted, and we can change the dependencies between words and add different grammatical information to bend and shape the meaning of the whole in myriad ways. No animal system is capable of anything remotely like this.

But there is more. Whereas in animal systems, even those of the great apes, the relationship between signal and meaning is as rigid as cast iron and entirely (or at least largely) innate, there is no connection between the sound pattern of a word and its meaning. Even for the apparent exceptions to this rule, known as onomatopoeia, the relationship is wafer-thin. Words like *wham!* and *zap!*, found mainly in comics, do capture some of the sounds themselves, but the link is fairly tenuous. Notwithstanding all the exclamation marks,

VROOOOAAARRRRRR!!!!! remains a meager reflection of the mighty roar of a racing car engine. Likewise, words like *tinkle, clip-clop, howl, yelp, bray,* and *bark* only vaguely resemble the sounds they point to. *Cock-a-doodle-doo* may sound pretty cocky to English ears, but the French know that roosters really go *cocorico,* the Dutch insist on hearing *kukeleku,* the Japanese are convinced it is *kokekokkoo,* and the Navajo hear *ha'ii'-ao.*

This lack of relationship between a word form and its meaning goes by the name *Saussurian arbitrariness,* after Ferdinand de Saussure, the Swiss linguist who first identified the phenomenon. The consequence of Saussurian arbitrariness is that to decide what a given sign means, we must first figure out which code—that is, which language—the speaker is using. Take the word *massa,* for instance. In English, this does not mean anything; it is simply not a word. But in Dutch, it refers to a large, unstructured quantity (you guessed it: a *mass*). To a Turk, it means table, to a speaker of Hausa it is a respectful form of address, like *Sir,* and in Bahasa Indonesia, it refers to time or an era.

This reveals how fundamental the distinction is between animal communication systems and human language, and how deep the chasm between them. The signals of animal systems are biologically determined. With very few exceptions, they are part and parcel of an animal's genetic endowment. The animal "knows" them as a matter of course, but cannot change them or add to the system in any way. In humans, however, all that is "given" is the basic plan or program that enables us to use language. This is what we are born with. But before this basic plan can mature into a fully operational command of one or more actual languages, children have a great deal to learn. First, they need to listen for the specific sounds of their mother tongue-to-be, uncover the rules that govern the formation of words, and pick up on the particulars of how sentences are put together and the ways in which the functional aspects of sentences are coded. This alone would suffice to explain why it takes us the better part of ten years, even without considering the second great task we face: acquiring a vocabulary, a gargantuan heap of well over ten thousand words even for the least educated among us, and in many people reaching fifty thousand items or more. Luckily, as children we are not aware of the enormity of the task, and even in retrospect we can scarcely imagine what actually went on. But when it comes to "learning words," the fact is that we all move mountains during the first few years of our lives. By the time the process reaches its peak of productivity, roughly around our third birthday, we are learning literally dozens of new words every day. And for most of us, the process never really stops: even if it is no more than in-company jargon, we keep on learning new words right up until our dying day.

Lastly, we must learn how to put all the rules and words we acquire to effective use. The rules of what to say to whom and when are complex, subtle, and largely unpredictable, so finding the right words and the right tone for the occasion is no easy matter. It is also almost entirely a cultural task, and thus never completely finished.

So although only a tiny part of human language is biologically determined, that itty-bitty part is of fundamental importance. Without the almost imperceptible but cast-iron scaffold known as universal grammar, we simply could not acquire language, let alone use it. Quantitatively speaking, however, by far the greater part of our linguistic knowledge is acquired through cultural transfer, throughout the course of our lives.

* * *

The very fact that the signals of animal communication systems are biologically determined and mutually unrelated reveals how such systems must have evolved: signal by signal, along the normal Darwinian path of mutation and selection. Systems like this, with no internal structure to speak of, are cheap in evolutionary terms. They are little more than a heap of signals, to which new signals are added as they emerge. What is more, virtually none of these signals arose for communicative purposes. Their communicative function is usually a freebie, an accidental by-product of a physical or behavioral characteristic that emerged and took hold in a species because it offered some unrelated evolutionary advantage. It is an *exaptation*, in the words of the legendary evolutionary biologist Stephen Jay Gould.

A good example is hair-raising, which signals acute stress in probably all homoeothermic animals. But the signal is only secondary. The real reason an animal raises its fur or ruffles up its feathers is energy preservation. When it is cold outside, this makes good physical sense. A lot more air fits in-between hairs or feathers that are standing on end than in sleek fur or plumage, making for better insulation. The effect is so advantageous that it can still be seen in humans, although we shed our fur long ago: when we get cold, we get goosebumps. But cold is just one of many types of stress our skin reacts to in this way. Anger and fright are others. And even sexual desire—think of what happens when your bare arm ever so slightly brushes against that of a desirable other: goosebumps!

Regardless of the cause of its state of arousal, a stressed animal is dangerous and unpredictable, especially if it has sharp claws or fangs. Animals able to recognize stress in others have greater chances of survival, and this is what gave hair on end its communicative function. A double function, in this case, because raised fur and ruffled feathers also make an animal look bigger, and thus even more formidable.

Take a moment to ponder how cleverly nature makes the most of simple means. An animal that conveys its tension by raising its hair does so without investing an ounce of energy in this communicative act. The signal is produced for completely independent reasons; it gets its communicative function only through the fact that other animals interpret it as conveying a message. These smart cookies gain an immediate advantage almost for free: the fact that they recognize the danger implied in the other animal's posture reduces their risk of having to fight or getting eaten. There is a profit in this for the agitated animal as well: rivals and predators will think twice before taking on such a fierce-looking opponent.

Something similar holds for maintaining the integrity of one's territory by means of scent. Excretion is essential in all organisms and so, like hair-raising, its communicative use is a bonus. Excrement has a bad reputation, and rightly so: it is made up of stuff a body has no use for. This is why we find poo and pee smelly—to encourage us to give it a wide berth. The more an animal's metabolism resembles our own, the stronger the effect. To us, dog doo is worse than rabbit droppings, but worst of all is the excrement of our fellow human beings. Other species feel the same way, and this is what makes excrement highly suitable for the purposes of demarcation. Excrement carries a built-in message that reads "Danger—hazardous substance alert." This is exactly the message the landed gentry of the animal world want to convey to would-be intruders.

* * *

As far as we are aware, all creatures in the animal kingdom share the same type of communicative system. To find exceptions, you do not want to look to our closest relatives. Look instead at humble insects, like bees with their waggle dance, or maybe ants, which appear to tap out messages on each other's heads. What exactly these ants are doing is largely still a mystery, but as we touched on earlier, we do know that even bee dances have an immutable, built-in meaning. Although they have some variables, these are restricted to expressing the direction, distance, and quality of the dancer's find. All other known animal systems are even less sophisticated. They are made up of a few unrelated, largely genetically determined signals that never vary and are never combined into more complex units. And though romantics will always point hopefully to the eerie songs of whales or the low-frequency stamping that helps elephants maintain contact over long distances, there is not a shred of evidence that they are any different.

Ramshackle and paltry as they may be, these systems seem to fit the needs of animal life to a T. There is not a single animal species around that has invested seriously in communication, let alone one that has come up with anything like human language. This is another strong indication that the evolution of language was not driven by communicative needs or advantages. It is hard to imagine why, among the millions of species that walk the earth, mankind alone should have experienced selectional pressure to upgrade its communication system, and in a very costly fashion to boot. It also puts paid to the idea that our capacity for language has developed from an earlier animal communication system, for several reasons.

First, there is the conservative nature of Nature itself. This can be seen in the strong biological similarities that still exist between ourselves and the tiny worms with which we parted ways hundreds of millions of years ago. Evolution is a tight-fisted tinkerer, always building on what is already there. It never throws anything out altogether only to come up with something from scratch. But as we have seen, the fundamental properties of our language faculty are completely at odds with those of animal communication systems. This suggests that they have unrelated origins.

Second is the simple fact that if language developed out of an earlier animal system, that system itself would no longer be present in us. Think of the common ancestor of humans and chimpanzees, which evolved into us on the one hand and modern chimps on the other. But this is not the case at all. In addition to human language, we still possess a communicative system with all the trimmings of a normal animal system, and a very extensive one at that. Utterances like "ouch," "psst!," "oi!," "humpf," and the tired "phhhhhh . . ." that escape us when we crash down on the couch at the end of a busy day are all part of it. So too is the collective "ooooooohhh!" that reverberates through every soccer stadium in the world when the ball misses the goal by a hair's breadth. "Hey, hey, hey!" and "yeah, yeah, yeah" and "whooaaaah" belong to it as well, as do things like laughing out loud, cheering, jeering and booing. Just like animal signals, all these cries and mutterings have set, unchangeable meanings and cannot be combined into larger meaningful units.

As you would expect, this animal system inside us also includes many signals that are not produced orally, but involve other modes of expression—making a face, for instance, or sticking out your tongue the way children do to express disdain or malice. Other examples are turning pale, blushing, breaking out in a cold sweat, fidgeting with uncertainty, jumping with joy, drumming your fingers and stamping your feet in impatience or anger. Waving can be considered one of these signals too. So can covering your eyes with your hands in panic or despair, and scratching your head or pulling your chin when you are suddenly at a loss for what to do or say.

Other signals involve several physical channels of expression simultaneously. Take the "beats me" look: hands spread wide, palms up, head tilted to the side, eyebrows arched, mouth pulled wide and lips closed. Or the shaky hand that follows a narrow escape, accompanied by a relieved "pfff" or even a soft whistle. And let us not forget the raised index finger accompanied by an admonishing "UH-uh" or "tsk-tsk." Better still is that complex signal typical of early teenage girls, consisting in a curt twist of the hips, as if they want to show you their buttocks, while they screw up their face, bow their head slightly, and blow a raspberry. Sometimes they will add a pat on the turned-out hip for good measure. Its meaning is something like "fat chance" or "no way, José, I'm not THAT stupid."

This system operates independently of both human language and culture, but does occasionally rub shoulders with both of them. What we usually refer to as body language is to a large degree but not completely culture independent. Certain elements, such as bowing, have become so ritualized in many cultures that their animal provenance is now long lost. The unworthy groveling of Gollum in *Lord of the Rings* may remind us of the behavior of a guilt-ridden dog, but it bears little resemblance to the demure, old-fashioned curtsy of well-bred German *Fräulein* called *Knicks*, the debutante's *reverence* or the highly formalized, carefully measured bows of the Japanese.

The animal system of communication in action. On her knees, with her shoulders hunched submissively, and a pleading expression, the woman needs no words to convey her anxious submission.

Certain oral signals have gone down a similar path. Some have almost acquired the status of proper words, including a stable spelling of their own. "Ouch" is perhaps the best-known example, but "oi" is up there too, as are "yuck" and "psst." Still, these are not really full-fledged words, since they cannot function in any grammatically meaningful position in an ordinary sentence. They only occur as interjections, tidbits scattered between the strands of the fabric of sentences—that is why we had to put them between quotation marks above. Unlike true words, they need not obey all the rules of a language's sound system, and they tend to resemble each other across all sorts of languages. The English "oi!," Dutch "hee!" (pronounced *hay*), and Spanish "eh!" (sounds like cockney *'ay*) are all locally colored versions of one and the same signal. This is exactly what we would expect of elements belonging to an animal communication system.

Oddly, things can also flow the other way. Whereas some animal signals have almost attained the status of true words, the old animal system seems to have been enriched by certain kinds of fossilized linguistic expressions. Instead of the originally animalistic 'ouch,' many a carpenter who hits his thumb instead of the nail will produce a heartfelt "goddammit!" Similarly, the humble "huh?" is often replaced by phrases like "what the hell?" or "Christ almighty." Just how intimately our animal and linguistic systems are intertwined I witnessed in the summer of 2007 in the southern Italian town of Lecce. Right before my eyes, an obviously well-educated, God-fearing motorist was rear-ended on a busy street corner. Before the eyes of a hundred bystanders, he slowly rose from his wrecked car, contemplated the damage with dangerously mounting tension, and then, arms wide, palms raised to the heavens, let out a heartrending.

CRISTOFÓÓÓRO COLÓMBO!!!

"Christopher Columbus": this was as close as he could bring himself to cursing. He could not let out a *porca miseria*—stinking bad luck—let alone a positively rude but, under the circumstances, eminently justifiable *porca Madonna*. It was a brilliant example of the civilizing effect of culture on wild, animal emotion.

Curses and oaths, or in BBC terms "strong language," all started out as purely linguistic phrases and formulas, then slowly solidified into unstructured cries of frustration just like the original signs of the old animal system. Oaths along the lines of "may God damn me if . . ." or "may God damn the thing that . . ." were honed into "goddamn" and "goddammit" as many as seven or eight centuries ago, and gained such popularity that Joan of Arc's contemporaries referred to their English adversaries as *les godams*. More recent examples are interjections like "(holy) shit!," "fuck!" and their more domesticated variants, all the way down to "jeepers creepers," "holy mackerel" and "drat!" Such stylized curses do not lose their linguistic status, just as true members of the animal system like "tsk" and "ouch" do not develop into full-fledged words. They never turn into something that violates the sound system of the language to which they belong, and "shit," "fuck," "good heavens," and the rest remain fully functional as nouns, verbs, adjectives, adverbs or phrases. Interestingly, only watered-down placeholders like "jeepers" and "drat," which are expressly coined to serve as polite variants, have no clear status. They do not originate in the animal system, but do not enjoy full lexical status either. That is, they are subject to the usual sound pattern rules, but cannot function as true parts of speech in a sentence.

* * *

If you still feel that human language arose from a need for better communication, two telling discrepancies between it and animal systems ought to give you pause.

One thing about animal communication systems is clear: they actually serve a communicative purpose in the strict sense. Passively or actively, they convey information. Animal systems are limited but effective, tailored to their task of spreading simple, unequivocal messages like "go away," "don't!," "you're okay," "I want to mount you," "danger, run!", and "there's food here." Such messages are either spread passively and continually throughout an animal's lifetime, as is the case with the colorful exteriors of the wasp and the pitohui, or only in reaction to an external stimulus—think of the Californian ground squirrel heating up its tail when threatened by a rattlesnake. The only exception is when animals deceive one another, as vervets occasionally do. As a rule, animals do not actively send messages unless there is a clear and concrete reason to do so. People use their animal communication system in exactly the same way. Except for sufferers of Tourette's syndrome, nobody shouts "ouch" or "yuck" for no reason. But human language is a completely different matter.

The first fundamental difference is that people do not need a reason to chatter away like mad. The only circumstance in which we prefer to keep silent in the presence of others is in the case of forced intimacy with strangers. Then, we carefully avoid eye contact and all other forms of acknowledging each other's existence. This happens in lifts, doctors' waiting rooms, and train compartments, and gosh, do we sometimes feel awkward there! In fact, we find collective silence such a burden that the "minute of silence" is among the most solemn rituals known to mankind, and one of the hardest to perform. Just how difficult was demonstrated by Sepp Blatter, the chairman of FIFA, the global federation of soccer associations, on 6 December 2013, the day after Nelson Mandela passed away. Blatter was presiding over the 2014 World Cup draw, a widely televised event. Before getting down to business, Blatter handed the microphone over to Brazilian president Dilma Rousseff, who asked for a minute's silence to honor the deceased South African statesman and anti-apartheid hero. Standing on the dais before the eyes of the world, Blatter managed to keep his mouth shut for all of eleven seconds.

The second point is that people use language in a great many noncommunicative situations. To begin with, we habitually "talk" to ourselves. That is, we all verbally accompany our daily activities with forms of inner speech. Sometimes we even do so out loud, talking to ourselves quite literally. This is something no animal will ever do. Moreover, it hardly counts as true communication; instead, it functions as a control mechanism, and a highly developed, multifaceted one at that. We talk ourselves through the process of preparing a fancy dish, or we count in our heads to regulate our rhythm and speed as we walk. And when we feel afraid or uneasy—when we happen to find ourselves entering a murky cave, a dark forest, a narrow path with a steep drop or a deserted house full of creaking floorboards—we all have a tendency to start talking to ourselves or even to nonexistent companions so as not to lose our nerve. All this is pure, unadulterated language, but no information is being conveyed to anyone.

There are other linguistic modes, too, that involve no real transfer of information. Think of the nonsense rhymes children are brought up with: "Eeny meeny miny moe, catch a tiger by the toe." What on earth is that supposed to mean? Not much, really. At least, this is what children think. No child is ever tempted to go out afterward looking for a tiger. Nor will any child enquire after the true significance of "miny" or "moe." Somehow they simply know it is nonsense, and are perfectly happy with that. Even as adults we indulge in huge amounts of nonsense, especially of the musical kind. Songs, whether they be classics like the Beatles' "She loves you, yeah, yeah, yeah" or momentary hits like Coldplay's "Every teardrop is a waterfall," rarely top the charts on account of their information content. Who cares that the girl from Ipanema goes walking, what a gadda-da-vida is or that the late Amy Winehouse said, "no, no, no" to rehab? In fact, song lyrics are almost completely irrelevant—often we cannot even hear what a singer is saying at all, especially when the language is not our own. Operas are notorious that way. And when it comes to one of the greatest pop classics of all time, the Rolling Stones' "Sympathy for the devil," there is only one fragment that everybody can sing along with flawlessly: "ooh-ooh!"

Crosswords are another case in point. Great masses of people apply themselves to solving them with no end of enthusiasm and diligence, bringing all their linguistic savvy to bear, and yet with exactly zero communicative intent or result. Crosswords are a profitable industry that serves people from all walks of life and many cultures, and they are usually filled out in silence and solitude. Or consider *Scrabble*™, the ultimate example of playing with words as mere forms, patterns of letters with numerical values but devoid of any real meaning.

None of the above activities has communicative value in anything but the vaguest, most general sense. When they happen to involve more than one person—when people solve a crossword puzzle together, for instance, or play *Scrabble*™—what goes on is not unlike what happens in bars. Language and conversation serve mainly as a social lubricant, contributing to an atmosphere of congenial well-being. We use them to make ourselves and each other feel we belong, are part of something. Few people hope to speak to nobody when they go to the pub. But the content of what is being said or sung is seldom paid attention to—and rightly so. It is small talk, innocuous chitchat that is not to be taken seriously.

That said, all we have managed to do so far is to deepen the mystery of the provenance of language. Because if not from our pre-existing animal communication system, where *did* it come from? And if its development was not driven by a need for better communication, then what set the process in motion, evolutionarily speaking, and kept it going? To answer these questions, we must first find out what conditions had to be met before our ancestors could even come close to developing a true language faculty.

3

Goodbye to the Animal Kingdom

"I can still feel those little hands," Marleen Janssen said in 2008 of the deaf and blind children she had been teaching for years. "You have to keep hold of them at all times. Otherwise, their world will simply stop." Janssen is the world's first professor of deafblindness. Try and imagine what it must be like: hardly or not at all being able to see *or* hear for as long as you can remember. Having no clue as to what the world looks like, or even what "looks like" means. Having never heard a sound. No voices, no music, no thunder, no rustling of leaves, no water lapping, not so much as a footfall or the tinkling of teacups. How do you cope with the world? Moreover, even in countries as highly developed and as affluent as ours, the unlucky can still find themselves branded plain mad and locked away in an institution. For those who manage to avoid this fate, life is still not easy.

And yet, provided they get adequate help, many deafblind people do cope more or less successfully. In spite of a lack of contact with the world around them, which the rest of us cannot begin to imagine, they learn to understand that world and come to terms with it. Some even manage to rise to great heights, like the renowned author and activist Helen Keller. Keller graduated from university, gave lectures, wrote books, and hobnobbed with intellectuals and presidents. Still, most of her world existed only inside her head. She knew the world in which she lived through taste, smell, and touch, but her most important lifeline—her sixth sense, so to speak—ran through the palm of her hand, where her lifelong companion, "teacher" Ann Sullivan, explained the whole universe to her through fingerspelling.

Keller was—and will remain—a rare exception. Perhaps the fact that she did not lose her sight and hearing until she was eighteen months old gave her a decisive head start. By that time, the development of her language faculty was already well underway, which gave her something that Sullivan's busy fingers could tap into and expand upon. Without that, it is hard to imagine how it would even dawn on you that a bit of scratching on your palm could actually be a message. Yet this does happen to deaf and blind people, albeit with great difficulty and often with only modest results.

This fact—that even congenitally deafblind people can hit upon the idea that certain tactile sensations are actually codes, messages that can be understood and answered—is one more piece of evidence that the core of our language faculty is innate, but not its details. Otherwise, we would never be able to switch to such an unusual channel. It also testifies to the sheer strength of our urge to develop these capabilities. The developing brain will seize upon anything it can in the effort to create a mother tongue. What is more, it shows just how important our inner world is—the whole jumble of experiences, memories, concepts, ideas, thoughts, feelings and drives that populate our brains.

* * *

If people who are both deaf and blind can develop into competent adults able to get by in the world, then direct contact with the external environment cannot be as indispensable as we usually think it is. And if that is true for the deafblind, it must be true for the rest of us too.

This same idea crops up when we observe the behavior of blind people with normal hearing. With a little help, blind people lead as satisfactory and fulfilling lives as everyone else. At different points in history, they have even enjoyed a highly prized status. The proverbial blind bard, the artist who depicts a world he cannot see, predates recorded history. Legend has it that Homer was blind, though nobody knows this for certain (or, for that matter, whether he even really existed). Be that as it may, his *Odyssey* pays homage to the blind "seer" Tiresias, who lives in the underworld. Or take the Roman goddess Lady Justice, symbolized by a blindfolded woman. Her very lack of sight serves to sharpen her insight, enabling her to dispense justice without regard for social rank or influence.

This is all just mythology, you might say. But a look at the simple facts of everyday life reveals something truly remarkable. Unless they mention it explicitly, there is nothing in the work of modern blind bards like Ray Charles, Roy Orbison, Stevie Wonder, Jeff Healey, or Andrea

Boccelli to suggest that they cannot see. Birds, clouds, mountains, the stars, the sky, the sea, and the sun: blind people seem to know all about them. Many even have strong opinions about the colors of things, and do not find that strange in any way. All this leaves us with just one conclusion. The concepts in our heads do not "represent" things in the external, physical world. They arise and exist in our minds almost completely independently of the world around us.

This notion is hard to grasp, but it is essential. We feel so strongly about being a part of the physical world, and we conceive so easily of words as "referring" to entities in that world, that we overlook the reality: concepts and the terms we use to label them do not derive their meaning from anything physically real.

Take the concept TREE, to which speakers of English refer by the word *tree* (while the French call it *arbre*, the Greeks *dentro*, the Turks *ağaç* [say a-h-atch] and the Chinese *chi* [say kai]). If this concept derives its meaning from some actual tree somewhere out there, which tree is it? Of course, there is no one tree that everyone in the world can inspect, so as to conclude, "Ah, so that's a tree / arbre / dentro / ağaç / chi!" Even if there were, trees die, and then what? Would the death of this archetypal tree cause the meaning of TREE to change into DEAD TREE? And what would happen once the busy little beetles had finished feasting on its rotting remains? Would TREE then cease to mean anything? Of course not.

If there cannot be one reference tree, could the word *tree* derive its meaning from all the trees in the world, collectively? Could it be that we understand the meaning of TREE only once we have seen enough of them? This cannot be true either. The problem is that you have to know what a tree is—and what is not a tree—before you can decide whether a particular tree-like object is part of the reference set or not. Otherwise, you end up with a hodgepodge of items that leads to all the wrong conclusions. It is a classic case of begging the question.

So the notion TREE cannot refer to one specific tree, nor to all trees. But neither can it derive its meaning from *some* trees, whichever subset you choose, for the same reasons. And, of course, it does not refer to anything in the physical world that is not a tree. That leaves only one possibility: the concept of TREE does not refer to anything in the outside world at all. Its meaning emerges and exists only within our minds. The world outside has precious little to do with it.

And so it is with all words and concepts: our inner world, with all its knowledge, feelings, and thoughts, is largely unconnected to the material world around us. From cradle to grave, it goes its own way, on its own terms, regardless of the world outside.

This is the point at which the philosopher leans back contentedly, and all ordinary people jump up to object: of course we learn by reference what a tree is, or a teapot, or a Bentley! We do so when somebody points one out to us and says, "Look, a tree / teapot / Bentley!" And whenever a toddler points to a bush and gleefully proclaims "tree!", grownups immediately set them straight: "No, Adelphina, that's a bush, it does not have a real trunk, see? Trees have *trunks*!" This may seem to be a valid objection, but think about what exactly creates or fleshes out the definition in the child's mind. It is not the thing itself, but rather the *idea* of the thing as conceived of by the adult. It is the parent *telling* little Adelphina that the thing she is pointing at is a tree—or a teapot or a Bentley—because that thing happens to conform to her *mental image* of those things.

There's no better illustration of this phenomenon than Top Gear's Jeremy Clarkson exclaiming, "Finally, the new Brooklands is a real Bentley again." A *real* Bentley? Since there are no fake Bentleys on the market, this can only mean that the Brooklands conforms to an *idea*, an abstract interplay of shapes, proportions, and materials that summons up an image in Clarkson's mind. The actual car he is talking about does no more than illustrate this idea and helps reinforce, perhaps even create, in viewers' minds the concept of a "real Bentley" as a special case of BENTLEY.[1]

What it comes down to is this. When people learn the meaning of a word, they do not forge a connection between something in the world outside and a notion in their mind. Instead, two minds become connected. A notion that was first present in one head arises in another head as well. Or, if that notion was already present in the receiving head, it is refined, extended, or corrected. The "thing out there," whether a physical entity, a photograph, or some other kind of representation, is no more than an illustration of a concept, at best a carrier for the knowledge transferred. And ultimately, it is superfluous. We do not even need to point at the real thing, because it is perfectly possible to explain to someone in words what a tree, a teapot or even a "real Bentley" is.

The same thing happens when, after our first encounter with a tree or a Bentley, we come across another. If we are not sure whether it fits the bill, we ask someone else, "Is that a Bentley?" The answer may be anything from "Duh!" or "That's a Mondeo, moron!" to "Well, sure, but only the nouveau riche drive *these* Bentleys—overpaid soap stars and soccer players, you know? It's not really a *real* Bentley." And so we tweak and tune our concept of BENTLEY, based not on the thing itself, but on the answers of the people we ask about it.

* * *

The gap between concepts in our minds and things in the physical world goes a long way toward explaining how blind people can "get" all sorts of typically visual concepts despite their lack of visual information. In many cases, simply being told about such things is enough. But that is not the whole story.

Next to all the things we "know" because we have been told about them, our minds host a set of concepts that do derive wholly or partly from sensations that reach us from outside. This is quite miraculous, considering the limitations of our senses. The magnetic forces that help migratory birds find their way around the globe, for instance, are completely beyond us. Except for the tiny fraction we call visible light, we barely notice the all-pervasive electromagnetic radiation that permeates our world. Unlike bees and many birds, we cannot even see ultraviolet light; our eyes are not sensitive enough to pick up its short wavelength. The only other part of the electromagnetic spectrum we are tuned into is infrared radiation. Although its wavelength is longer than our eyes can see, we can feel its heat through our sense of touch. The rest of the electromagnetic spectrum we experience only indirectly. UV light darkens our skin; an overdose can result in sunburn or even skin cancer. Radioactive radiation will kill us, but we have no way of recognizing it or its pernicious workings until it is too late. And even then, it is not the radiation we notice, but the pain we feel as our bodies disintegrate.

When it comes to hearing, humans are mere bunglers. Even the dimmest pet hears frequencies far higher than the highest sounds that human ears can pick up. To make matters worse, our hearing begins to deteriorate as soon as we reach adulthood. The bandwidth of audible sound narrows so fast that some Dutch municipalities have installed systems for dispersing groups of loitering adolescents by means of a nerve-racking, high-pitched whistle people over 25 can no longer hear. At the other end of the spectrum, large animals like elephants and whales can hear sounds so low that we experience them only as vibrations in our bellies.

The same goes for smell, taste, and touch: the human senses are mediocre at best. Compared to other species, it is fair to say that even perfectly healthy people stumble through the physical world half deaf and blind. And even then, the impressions that do find their way to our brains are only vaguely connected to the world outside. All that ultimately reaches our brains are electrochemical pulses generated by excited nerve endings, with vast amounts of information already filtered out.

A good illustration is the way we experience the sounds of language. No two voices are truly identical, and even the same voice does not always sound the same. People skip over sounds and cut off the ends of words. They mumble, they lisp, they yell over ambient noise and they wheeze through a cold. As a consequence, the raw speech sounds that reach our ears vary so wildly that a visual representation reveals nothing whatsoever about what is being said. Try running some speech through a sound editing program: you cannot even see whether a given sound is part of this word, that word or is not a word at all. Yet we have no trouble identifying the roughly forty to sixty speech sounds of our own language, even in noisy conditions.

Let us be more precise. The difference between the vowels in *car*, *race*, and *beer*, for example, has to do with how far the blade of the tongue is raised toward the palate. In *car*, the tongue is held low in the mouth, in *beer* it is raised all the way up, and in *race* it is somewhere in the middle. In reality, the sound characteristics—the sound of the sound, so to speak—change gradually as the tongue moves from one position to another. This glide from one sound to another is known as *glissando*. Think of the way the pitch of a violin changes as the violinist's fingers slide up and down a string, one sound melting smoothly into the next. But what we *hear* is more like the discrete sounds of a guitar. Thanks to its frets, the guitar, like speech, has no "in-between" sound. Go up one fret, and the pitch changes abruptly from B to C.

In speech, as we raise our tongue, the *a* sound in *car* turns suddenly into the *a* sound in *race*. And as the tongue keeps going up, it just as suddenly becomes the *e* of *beer*. This phenomenon is called categorical perception: what we perceive is a digitized version of the actual sounds, in which the full gamut of auditory qualities is reduced and compartmentalized into discrete units. This is a nifty ploy that helps us interpret speech correctly at high speeds, in noisy conditions, and independent of the individual qualities of a particular voice. But it is a long way from what the physical world actually presents to our ears.

Somehow, our auditory system knows which of the auditory impulses we receive must be considered speech-like, and preprocesses them accordingly. We are not aware of this going on, nor do we have any say in the matter. Try as you may, you will never hear the glissando your mouth produces as you go from *car* to *beer*.

In categorical perception, the only properties of sounds that count are those that discriminate between elements of a language. All others are discarded as noise. For vowels like *a* and *e*, pitch is relevant, but not friction. This is why we can understand the vowels in *car* and *beer* equally well even when we have a cold.[2] For the consonantal trio *f*, *th*, and *s*, the defining characteristic is the precise kind of friction. We can pick up without fail the subtle differences between friction between our lower lip and upper teeth (*f*), the tip of our tongue and upper teeth (*th*), and the front part of the top of our tongue and the palate just behind our front teeth (*s*).

Categorical perception is a high-level skill. Like other high-level skills, you do not get it completely for free. It has to be tuned, and this is why it is so hard to make sense of unfamiliar languages when we hear them for the first time. Think of the first time you saw an Italian film, or heard someone recite a poem in Georgian. You can barely pinpoint the consonants and vowels used, let alone identify the boundaries between words. To our brains, unfamiliar languages are just a flow of sound, more like music than speech. Unlike the acquisition of language per se, however, categorical perception is not bound to a critical period. You can learn to process the sounds of Turkish, Chinese or Cherkessian at seventy just as well as you can at seven.

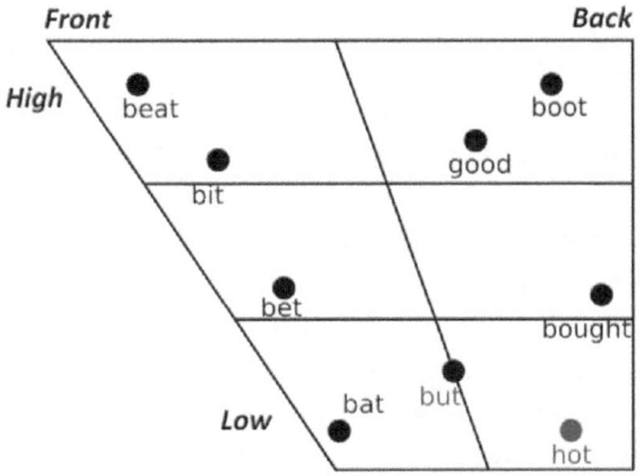

The tongue's position in the mouth when producing the main English vowels.

A second characteristic of categorical perception is that it is not limited to sound. People who "speak" sign languages use categorical perception too: the visual signals they receive are preprocessed and packed to neutralize the endless variation in gestures caused by differences in the size and build of the signer, and their sloppiness, exuberance, or agitation.

Meanwhile, we are neither conscious nor in control of our categorical perception. We have no power over it, nor can we shut it down. Even knowing about it on an intellectual level does not enable us to hear the glissando from *car* to *beer* that is actually there.

Like the language faculty itself, categorical perception is one of those partly hardwired, biological capacities of the brain, and as such it can be impaired at birth or damaged in the course of life. There are indications that dyslexia—difficulty in recognizing and reproducing words, especially when written down—is caused by impaired categorical perception. Dyslectics, simply put, hear too much, so much so that they cannot see the forest for the trees. As we know from sign languages, the brain circuits involved in preprocessing the sounds of speech are involved in visual tasks like reading and writing as well. And so dyslectics make mistakes in reading, writing, and hearing, but have no trouble speaking. The reason for this may be that speech production does not have to be all that precise. The hearer's categorical perception will see to it that whatever you blurt out is reduced to a standard, recognizable sound pattern. This does not work with written text, of course. In that case, you really do need to dot the i's and cross the t's.

And so we come to the prince of our senses: sight. We think that what we see is what our eyes observe, but the truth is that seeing involves even more preprocessing, filtering, and censoring than hearing. The coherent flow of scenes that we experience, like something out of a film, is a far cry from the raw material gathered from the retina. The processes of sifting, sorting, and preprocessing kick in right there at the retina. By the time we become aware of anything at all, a lot has happened already. Impressions have been corrected for eye movements. Visual chunks have been prised apart into elementary shapes and colors, analyzed and put together again, cleared of noise and irrelevant detail, and integrated seamlessly into the flow of time. All this goes on, without pause, for every moment our eyes are open. It dwarfs the accomplishments of Hollywood and all its clever machinery.

What ultimately reaches our consciousness has been cleaned up and reconstructed so thoroughly that eyewitnesses are notoriously unreliable, even before time has had the chance to play tricks with our long-term memory. In one experiment, half the subjects watching a basketball match on television failed to remember or even notice the man in a gorilla suit who ran right across the court. Likewise, to the frustration of police, people who see an accident happen right before their eyes often register very little of what went on.

* * *

The reassuringly clear and stable image of the world our senses serve up is more than hopelessly incomplete. Thanks to all that sifting and streamlining, it is very indirect as well. It is not unreasonable to conclude that we have very little true contact with the material world at all. How we behave and survive in the world depends solely on patterns of neurons in different states of agitation inside the brain. These patterns are connected to and command our bodies, to some extent accidentally, in such a way that we usually interact with our environment quite adequately.

This constantly changing set of patterns is our inner world. It consists largely of more or less fixed concepts, linked by permanent and temporary neural connections. And this mishmash of concepts is complemented by an ever-growing (and eroding) mountain of memories, plus all sorts of signals that are constantly arriving at the brain from all over the body. Some of these signals, the ones we know as sensory input, originate in the external world. Others emerge from somewhere within us, sometimes even in the brain itself—information about how we feel, what our hormones are up to and so on. It is the interaction between all these that determines our behavior, our conscious experience, and our sense of well-being.

The implications of all this may seem odd. Think of an acrobat on a trapeze. Even this routine—minutely orchestrated, flawlessy executed—depends upon the essentially coincidental correspondence between the real world and the neuronal state of his brain, which operates his muscles. But remember our flight simulator. Simulation software, too, has no contact at all with the outside world, and yet it behaves perfectly adequately with respect to that world, even reacting with admirable precision to the sensory impulses of a user operating the controls. Both the sofware of a flight simulator and our brains are based on a model of the outside world, without physically resembling, being part of or being in direct contact with that world.

Flight simulators and our inner world differ mostly in the way they have come into being. Flight simulation software is man-made, purposefully put together bit by bit by a carefully assembled team of specialists according to a predetermined set of functional demands. Its development begins with a detailed plan, a blueprint that is then executed step by meticulous step. There may be minor modifications along the way, but the plan essentially remains unchanged.

Our inner world is different. As a product of evolution, it acquired its form and functions with no plan, no engineers or programmers. Everything depended on what random steps evolution took and in what order. Now and then this aimless evolutionary mucking about resulted in something viable and useful, leaving us, in hindsight, with the illusion of a plan—a distinct impression that a particular route was travelled on purpose to arrive at a clearly envisaged goal.

* * *

This is the festering sore of the theory of evolution: it is very hard to accept that the whole elegant, efficient machinery of life is the fruit of brutal coincidence. Many people have trouble swallowing this even for the humble flagellum, a tail-like appendage that enables certain kinds of bacteria to "swim," not to mention complex phenomena like our sense of identity, consciousness, memory, and capacity to learn. Those unable to wrap their heads around this harsh fact often fall prey to two popular but pernicious misconceptions.

The first is the idea that creating something complex, clever, and beautiful requires careful planning and meticulous craftsmanship. But this is only our own experience as *homo faber*—man the maker—talking. Ill-conceived, sloppy tinkering will get us nowhere. What we need is creative thinking, intelligent planning, and perseverance, combined with a pinch of genius. We like to think that we always go about making, baking, or building complex and important things in a rational, orderly way. Hence the havoc that resulted when the renowned Dutch painter Karel Appel revealed the secret behind his highly controversial genius: "I just fuck about a lot." With these flippant words, he turned one of our most deeply rooted, generally accepted convictions on its head: the belief that the amount of physical and mental energy we expend on something is directly tied to its ultimate value. Even more infuriatingly, many could not help realizing that Appel's point, though painful, was valid. Now and again, it is possible to create something valuable and beautiful without really trying. Conversely, no amount of diligence will infallibly lead to a worthwhile result.

The trouble is this. Our opinions and convictions carry no weight beyond the confines of our bodies. Yet in trying to understand the world at large, our own human experience is all we have to go on. We simply cannot help anthropomorphizing. When we try to imagine what life is like for our dog, from the dog's perspective, we automatically turn the dog into a little human. We assume that it sees life basically like we humans would. We imagine it to think like we do and to feel the things we do, simply because we truly cannot imagine anything else. If you have doubts about this, just try to picture what it would be like to be a brick in a wall. See yourself snugly cemented in place, looking out at passers-by, fully conscious and aware. Putting yourself in the position of a dumb inert brick is as impossible as imagining yourself dead. But nature does not care what we can and cannot imagine about how it operates. It just goes its own way.

The second mistake concerns chance. By and large, we think of chance as a lottery with a great many tickets, most of them worthless. There is only one first prize. And although the chance of winning it is infinitesimally small, this is what we tend to focus on. It is all or nothing—and nothing is what we usually end up with. We cannot help but think the same way about the chances of such an immensely complex world as ours arising purely by accident out of pre–Big Bang nothingness. Surely these, too, are so minute as to be virtually impossible.

There are two flaws in this line of reasoning. First, we tend to overlook the fact that the chances of the prize falling on *any* particular number in a lottery are equally small. Imagine a scratch card lottery in which each card has a different number, and you can name a winning number before you scratch your card. It does not matter what number you pick or which card you bought: the probability of that particular number appearing on your particular card will always be one divided by the total number of cards on sale. All other things being equal, this is a general truth. If there are a great many possible states of affairs and only one can actually materialize, then all outcomes are equally likely—or unlikely, if you will. And yet one of them *will* become reality, just as there is always *someone* who wins the lottery. One-in-a-million chances come true every day.

Second, we should not view evolutionary chance as a simple matter of all-or-nothing. The world was not created in one fell swoop, just like an ape with a paintbrush will never produce a perfect replica of Sandro Botticelli's *Venus* in a single broad stroke. That is the stuff of creation myths. One fine morning a supreme being gets out of bed and—presto!—creates a complete universe, trimmings and all, out of thin air. Evolution is not like this. It is more like Karel Appel: it just "fucks about a lot," taking random, tiny steps through the ever-changing space of possibilities thrown up by circumstance. Most of these steps, which take the form of mutations in the genome, come to nothing. The few mutations that do make a difference often have only small, gradual effects. And very rarely, one such step turns out in hindsight to have kicked off a major, perhaps even a fundamental change.

Most importantly, these steps are completely independent of each other, just like the outcomes of rolling a die over and over again. Take the development of the humble amoeba into creatures as diverse as the orangutan and the octopus, over the course of more than two billion years. If we view this development as a single, linear process, it seems as wildly improbable as throwing millions of consecutive sixes. But this is the wrong way to look at it. A mutation in an orang-utan today has nothing to do with a change in an amoeba billions of years ago, nor with any other previous state of the organism we now know as the orang-utan. All that matters is the immediate effect of a mutation on the organism as is, in the prevailing circumstances of the moment. Evolution is always about one tiny step, one mutation at a time. It is not looking ahead to the future, taking scheduled steps toward some predetermined goal.

That being said, individual mutations that do "improve" a species are not terribly unlikely, although some animals are so well adapted to their habitats that they have remained unchanged for millions of years—flies, for instance, and cockroaches. So successful and robust are these species that the chances of fundamental improvements are next to zero. They continue to mutate, of course, but these mutations are unable to take hold for want of an advantageous effect. Less sturdy species, like Charles Darwin's finches, will more readily change in appearance and behavior under environmental pressure. If at some critical moment there are no adequately adapted mutants around, the species will go extinct. This too happens on a regular basis.

The challenge for biologists is to discover what particular steps were taken along the way from one form to the next. From amoeba to orang-utan, or from early reptile to modern bird. Or—somewhat less ambitiously—from a single-cell organism to an organism with a flagellum. Or even the other way round, which can happen too. With each step, all that matters is what changed and how that created an immediate selectional advantage for the bearer. In other words, how did it affect the bearer's chances of procreating and thereby perpetuating its genome? We are still in the dark about most of this. However, not knowing exactly how a particular developmental trajectory came about does not negate the facts of mutation and selection. Remember that less than 150 years ago, people had no inkling that viruses were responsible for the spread of certain infectious diseases. Yet that did not stop them from spreading, with sometimes devastating effects. Nature does not care what we do and do not understand, any more than it cares about our ideas and values.

* * *

There is nothing particularly human about having an inner world. Animals have inner worlds too, with varying levels of complexity. We can infer this from the fact that animals can learn, remember things and behave in consistent ways—even creatures as unassuming as *Aplysia*, a sea slug more commonly known as the California sea hare.

Aplysia can grow up to 2.5 feet long. Despite their impressive size, they have an extremely simple nervous system, with a few thousand neurons that are large and easy to spot. It was thanks to these qualities that they came to the attention of neuropsychiatrist Eric Kandel in the early 1970s. Kandel was interested in the manipulation of synapses: the sites at which two neurons all but touch. There, if the conditions are right, a kind of spark forms, and through this "firing," neurons communicate with and influence their neighbours. Information can spread in this way through entire networks of neurons, ultimately resulting in observable behavior.

The big question for Kandel was whether or not the properties of particular synapses could be changed. If a synapse's firing threshold, so to speak, was fixed for life, every brain and every nervous system would always respond to a given stimulus in exactly the same way. But if the properties of synapses were malleable, Kandel figured, it might explain how individual creatures are able to acquire their own, novel memories and display individual learning behavior.

To prove his idea, Kandel embarked on a series of experiments that eventually earned him a Nobel Prize. He first demonstrated that *Aplysia* were capable of learning. When exposed to a weak electrical

Aplisya, the California sea hare.

shock, they would retract their vulnerable extremities: their gill and their siphon, a sort of snorkel they use to draw air from above the surface of the water. After repeated shocks, they began to react faster and faster. What is more, when Kandel exposed them to a gentle "pulslet" beforehand, they even learned to retract their gill and siphon in anticipation of the real shock. However,, when these pulslets were not consistently followed by the real thing, the sea slugs quickly learned to ignore them. And slugs that were left to their own devices before being put back in the programme, say three weeks later, seemed to pick up where they left off—they had retained the memory. And Kandel had been right about how memories come about: the changes in behavior of the slugs corresponded to changes in the electrochemical properties of their synapses. Since the structures Aplysia use to learn are found in all species, having an inner world is obviously the norm.

The inner world of *Aplysia,* such as it is, is actually larger than its modest capacities for learning suggest. Swimming, resting, mating, and whatever other activities a sea slug engages in are all preprogrammed into those few thousand neurons. And more complex species are no different. From the days of René Descartes, the scientific consensus on the mental make-up of animals was that they act and react by instinct only. Unlike humans, they lack free will and consciousness, not to mention a soul. They are mere automata, driven by inborn, immutable instincts that control their reactions to the world in which they live down to the last detail.

Nowadays, ideas about the animal mind are somewhat less mechanistic. Yet there is still some truth to the Cartesian view. Young animals, for example, are surprisingly capable of fending for themselves. Think of a newborn calf. It may be shaky on its legs for the first few hours, but it already has a whole repertoire of behaviors at its disposal. As it cannot possibly *learn* these behaviors in so short a time, they must reflect an innate inner world. Taking its first wobbly steps, shadowing its mother, finding her udder, figuring out how to suck her teats, taking cover when danger looms—if a calf is to survive, all these skills are indispensible from the very first minute.

These are more or less mindless automatic motor skills. But the innate behavioral repertoire of calves also involves real concepts and real knowledge of the world. Recent reseach shows that young cows suffer from fear of heights. They have poor eyesight, and their eyes, on the sides of their skulls, are incapable of stereoscopic vision, which is what helps us humans perceive depth. Yet, they have no problem spotting holes and understanding that they ought to give them a wide berth. The modern milking stable often resembles a car repair shop with an old-fashioned grease pit: the cows are parked side to side in rows, their behinds poised over a long pit. Down in the pit, the farmer can hook up the cows to a milking machine without having to bend over for each animal, just as mechanics used to drain spent oil from a car. The research showed that when young cows that had never been milked were driven through such a stables, they became quite nervous if the pit was open for them to see. Even after they had become quite familiar to the place, their heart rates still shot up each time they passed the pit, unless it was covered.

It is hard to see how these cows could have "learned" to distrust heights. Dairy cows usually inhabit the reasonably flat world of the farm, where there are few cliffs or chasms that might inspire a fear of heights. The young of other species, too, seem to know from day one that heights are dangerous and falls are best avoided. Studies show that this holds for kittens, puppies, lambs, piglets, young rats, goats, monkeys, and even human infants. Place them on a raised plate of glass, half of which is see-through and half painted over, and they all wriggle onto the "safe," opaque part of the glass as fast as they can.

So this is one more similarity between humans and other mammals. And if we dig a bit deeper, we find that "height," "fall," and "danger" are not the only innate notions we all share. Have you ever needed someone to explain to you what a pit is? What about a hole, a mountain or a tree (yes, that too)? The sun, the moon, grass, earth, fire, water, a wave, a forest or a path? Or heat and cold? Of course, you had to learn the words for all these things, but not, in all probability, the concepts themselves. When was the last time you saw a child point at a puddle or a pile of sand and ask, "Daddy, what's that?" Even the sea is something kids seem to just *get*. Parents rarely need to explain what such things are. They just point out those hazards that are not immediately obvious: drowning in a dangerous current, say, or suffocating when a tunnel caves in.

* * *

All these innate concepts have to do with the typical, basic features of our natural environment. They all play a prominent role in the lives of larger mammals, like us, and we all know how to handle them with no effort whatsoever. We could postulate that human infants somehow pick up these concepts from their surroundings during the long period in which they are otherwise helpless. But for most animal species, this is not an option. In a matter of hours, their young are quite independent. Without some innate conceptual dowry this would be impossible. And if they have one, chances are that we do too.

Other concepts highly likely to be innate are those related to phobias. A phobia is an irrational, uncontrollable fear with no indentifiable cause. As far as anyone can tell, these fears are not rooted in some concrete negative experience, like when you have a car accident and can no longer bring yourself to get behind the wheel. Real phobias are not linked to any chance event. In fact, they are not random at all, but always about the same kinds of things. The usual suspects are heights, darkness, spiders, snakes, dogs, rats, and insects. Others include open spaces or cramped quarters, large bodies of water, thunder and lightning, vomit, excrement, blood, needles, pain, and so on.

The fact that such fears lack an external trigger suggests that they originate in our inner world. This is corroborated by the fact that common phobias occur with roughly equal frequency all over the world; culture does not really come into it. Thirdly, phobias are almost invariably about things that are also important in the animal world, just like other innate concepts. Most phobias have to do with things in our natural environment that might present real, physical dangers. A small dose of agoraphobia may be no bad thing, if it translates into paying more attention to potential hiding places. An aversion to water is a must for inept swimmers. It just makes sense to give creepy-crawlies and large beasts a wide berth. Vomit is clear proof that something is wrong with you or someone around you, and excrement presents all kinds of health hazards. Being mildly wary of all these things makes perfect evolutionary sense. If you stay out of the way of snakes and do not stray too close to cliffs, chances are you will live longer. It is only when the fear aspect of a concept gets out of hand that a real phobia arises.

However, phobias are never about things unrelated to our natural environment. There are no phobias arising from products of the human mind, such as birthdays, taxes or Tesco, however deep our aversion to these may be. There is no such thing as soprano-phobia or justice-phobia. People may be scared of lifts—the safest of all modes of transport—but this is really just a mixture of claustrophobia and fear of heights. Our aversion to the dentist has natural causes too: we dread both pain and being left defenseless in a vulnerable position. Even obsessive hoarding, avarice, and overeating are no more than misdirected forms of our evolutionarily quite sensible fear of hunger.

Phobias, then, stem only from concepts in our primeval environment. They are neither triggered externally nor do they differ from one culture to the next. So it is highly unlikely that we *acquire* phobias. On some level, they must be innate.

* * *

So far, all the concepts we have posited as innate have been concrete aspects of the natural environment, physical sensations or emotions. Concepts like trees, spatial notions like height, sensations like heat or agression. But at least one class of *abstract* concepts seems to be inborn too: numbers.

There are quite a few animal species that can do a little counting. Hedgehogs, for example, see the world in terms of "one," "two" or "lots." Remove one of its offspring from a nest of more than three and a hedgehog mother will not notice a thing. But if only one or two hedgehoglets remain, she will be perturbed. And many other species are similar: they can distinguish between small quantities, like one, two, and sometimes three, but larger numbers leave them in the dark. They may be able to make rough estimates, or differentiate between a bowl with five goodies and, say, fifty. But broadly speaking, anything more than a few is simply "a lot."

People are very different in this respect. In principle, we can keep on counting right up to inifinity. Everybody knows that 23 comes before 24, just as 90,635,692 is followed by 90,635,693, 999,999,999,999 by one trillion and so on. But that is not to say we can actually grasp such large numbers. We do not have a clue how much 90,635,692 really is. Would 90,635,692 bricks fit in a lorry? What about 90,635,692 sugar cubes, or 90,635,692 grains of salt?[3] We can barely read such large numbers, let alone spot the considerable difference between 90,635,692 and, say, 99,635,692 sugar cubes—or even 74 and 86—without a lot of painstaking counting. We know this full well, because our weak grasp of numerical quantities is the lifeblood of crooks and swindlers.

Determining the exact number of a great many items is a purely linguistic operation. Counting is just reciting numerals in the right order, beginning with one, and ticking off a unit each time. A large number literally equals the last numeral recited, no more, no less. It has no deeper meaning in our minds than that. And if we make a mistake along the way, all we can do is start all over again.[4]

Small numbers are different: we can really "see" them. We do not need to stop and count to know whether there are two, three or five plates on a table. Such small numbers are separate concepts in themselves. Only when we get to around nine do we need arithmetic to get to grips with numbers. Up to twenty or so we can still estimate reasonably well, but beyond that our guesses become wildly unreliable. If the police report that two to three thousand protesters took to the town square, activists are bound to have seen a good eight thousand. Assessing the size of a large crowd is as hard as guessing the number of beans in a pot—maybe even more so.

Another clue that we have no real conception of numbers beyond nine or ten is that the great majority of languages use the decimal system, with its ten units. Duodecimal systems are not unheard of: the ancient Sumerians used one, and the day still has two times twelve hours. But today these base 12 systems are all but extinct. At the other end of the scale, the binary systems used by computers are also exceedingly hard for the human brain. This suggests that a number system with which we feel comfortable must have approximately as many basic units as there are true number concepts in the brain.

Of course, there are always exceptions. One of the more bizarre number systems is the base 27 scale used by the Oksapmin and Foe people of New Guinea. The units are derived from parts of the upper body. They count not only on their fingers but also all the way up their arms, shoulders and head, including their eyes and nose.

The concepts of small numbers seem to be innate to humans, likewise the ability to make roughly correct estimates of somewhat larger amounts. That is, these things are built into our brains, not acquired or derived through language. In remote parts of Australia, there are tribes whose languages are virtually devoid of number words, and who have no knowledge of the rules of arithmetic. Counting is not really on their radar—but as it turns out, this does not affect their ability to make reasonable estimates about quantities. In 2008, researchers gave three groups of children a number of arithmetic tasks that required no words. One group consisted of English-speaking children from Melbourne. The other two were made up of monolingual speakers of Warlpiri and Amindilyakwa, Aboriginal languages spoken in different parts of Australia. Broadly speaking, these languages only have words for "one," "two" and "a large number" (Amindilyakwa also has a word for "three").

The children were seated in front of two mats. On one of the mats lay a number of items (never more than nine). First, the children were asked to put on the other, empty mat the same number of things that they could see on the first mat. All three groups turned out to perform roughly equally well. Later, they were asked to do the same, but this time the researchers covered the first mat with a cloth and added a few extra items while the children looked on. Again, there were no major differences in performance between groups. What this shows is that the Aboriginal children were perfectly capable of imagining quantities from one to nine without the benefit of number words. Not only that, but they proved able to divide a number of teddy bears equally between them—no mean feat for children of any stripe.

So to identify small, discrete quantities, we do not need to know the words for numbers. It is a skill we possess innately, regardless of the language we speak. But it is like language in the sense that only the general idea is present from scratch. The particulars are acquired over the first year of life. From the age of about six months, babies start to notice the difference between groups of two and three items. But the difference between four and six is still beyond them. The concept of "four" does not seem to materialize until the age of about ten months. This is long before the first word comes along, let alone the first number word.

* * *

Innate concepts are not related to the world outside—not any more at least. Perhaps they are best thought of as permanently agitated neural circuits, each of which more or less accidentally originated long ago in some ancestor of modern man. Through ordinary natural selection, they spread from a single individual across the entire species and influenced the behavior of their bearers in a way that enhanced their chances of survival.

This is how instinctive behavioral patterns must have developed, both in animals and in us. But what about all those innate concepts that simply do not tally with the physical world we inhabit?

One of these counterfactual concepts is the idea that we live in a stable, stationary environment. A world in which the sun rises in the morning and sets at the end of the day. What could be less controversial than that? And yet, our mind is playing tricks on us. In reality, the world is a huge clump of dirt and iron spinning madly around its own axis, shooting through the universe at breakneck speed. You think you are leisurely sipping that cocktail on a Bali beach? Think again. The truth is that you are hurtling around the Earth's axis at more than a thousand miles per hour. And those sunsets you so treasure? The sun has precious little to do with it. It neither rises nor sets. What you are seeing is the Earth turning and hiding it from view. The rest is an illusion. Rainbows are not really there. Roads do not narrow with distance, train tracks do not meet on the horizon, and cars do not grow larger on approach. One way or another, these are all illusions. We know they are not "real," and yet they persist—to our advantage. We have lived by them since the beginning of time. Our entire existence is structured around the phases of the moon and the path of the sun through the sky.

What must all this be like for animals? Chimps, say, or horses? Do they see the sun rise in the morning? Do they experience it as going down at dusk? The eyes of primates and other mammals are not unlike our own, and their senses are essentially the same. We can assume that they are familiar with rainbows and the tricks of geographic perspective. But is not the sunrise just a phenomenon created by our minds? After all, in nature, the sun does not "rise" in the way that we conceive of it. It just gets light out. Even if our animal cousins do see it in the same way, we will never know. They cannot tell us about it.

* * *

"Hm . . .," said Chief Inspector Barnaby in an episode of the BBC series *Midsomer Murders*, "Perhaps he has found the legendary Haddington emeralds."

"Do they really exist, then?" his wife asked, incredulous.

"Of course not. That's why they're legendary."

With these words, the television sleuth reveals exactly where the animal kingdom leaves off and the mind of man begins. For the vast expanse of our inner world is home to all sorts of notions with no relation to the real world at all. These are concepts that do not, and often could not, exist in the outside world. They are pure figments of the mind. As such, they cannot exist for a single second longer than we do. When the last human being on earth dies, each and every one of them will have vanished without a trace.

Let me give you an example. A few moments ago, those nonexistent Haddington emeralds, which you had never heard of before and will never hear of again, took up residence in your inner world. Already they have a whole host of properties, which you attributed to them by associating them with other elements of your inner world. Within that mental realm, even their very fictitiousness is just a property like any other, a property that cannot exist in the physical world. Out there, nonexistence is not just another kind of existence. In the physical universe of Newton and Einstein, things that do not exist have no properties and exert no influence. Our inner world, however, is rife with nonexistent things with all kinds of properties and powers. As Barnaby suggests, it is precisely the emeralds' lack of existence that gives them their fame.

Children mix things that exist in their inner and outer worlds in spectacular ways. Most kids go through a phase in which they drag a teddy bear or rag doll around with them everywhere they go. To them, their cuddly soulmate has a full-fledged personality. Teddy must eat with the rest of the family, watch television, and go to bed when they do. Children talk to their toy and treat it like a living thing—and they expect their parents and siblings to do the same. That it is made of cloth with a few lopsided beads sewn on does not matter. Its essence is a product of the child's inner world. Little Johnny does not understand that, outside his head, products of the inner world can only exist as words.

Imaginary friends are similar, but without a physical placeholder. Most kids just chat with their imaginary friend. But some go further. They may lay claim to a specific chair, for instance, that only their imaginary friend is allowed to sit in. Have you ever been accused of squashing your child's invisible friend by sitting in his seat at the dinner table?

Adults are just as fond as children of products of our inner world that cannot exist in the real world, if not more so. They are perfectly happy to accept the idea of a foul-mouthed duck in a sailor's shirt with white-gloved hands instead of wings. Donald Duck is even an honorary alumnus of the University of Oregon. Members of Mensa, the international high IQ society, dote on such unlikely characters as SpongeBob SquarePants and Homer Simpson. Or take A. A. Milne's talking teddy bear, Winnie the Pooh, whose honey-glazed quasi-philosophical musings are every bit as real as Pooh's friend Christopher Robin—who actuallly *did* exist. Few of Milne's readers know about that, and certainly none care. That a great many of the things inhabiting our inner world do not and could not exist outside of it is simply of no consequence.

It would also be wrong to think that products purely of our inner world are restricted to the realms of entertainment and fantasy. We busy ourselves with them in real earnest each and every day. Think of historical figures and other things that once enjoyed a physical existence, but no longer do. The Code of Hammurabi, the Ark of the Covenant, the battlefield at Alesia, the great library of Alexandria, the Colossus of Rhodes, Genghis Khan, the Ming Dynasty, the Court of Versailles, the Third Reich. They and countless others have all vanished from the face of the earth, some of them without leaving a trace. Yet in many minds, they are present and very much alive. Then there are a great many notions that can only have sprung from our inner world. These include notions like "democracy," "economy," "Peter," "garden," "monument," "itinerary," "yesterday," "beautiful," "insufficient," "table manners," "fast," "god" and "meaning." We know them only by the properties we ourselves have attributed to them, or through learning about them from others.

Together with historical figures, concepts like "democracy," "yesterday," and "meaning" belong to the class of notions we can only acquire by means of language, unlike "stone," "tree," or "car." There is nothing in the physical world pertaining to "democracy," nothing to observe, nothing for anyone to point at. We only learn of its existence when we are told by others, and we can only identify it by its name or a description. This indicates that language is a prerequisite for the development of at least *these* parts of our inner world. And that, in turn, suggests that they are the property of mankind alone, to the exclusion of all other species.

Things are slightly different when it comes to notions like "Peter," "garden," and "monument." "Peter" is usually made of flesh and blood, and is very much a physical entity, be he a man or a marmoset. One can point out Peter, a garden or a monument. Nevertheless, the meaning we give to such notions is completely lost on animals. To them, a garden is a stretch of nature like any other: a random assembly of trees, shrubs, grassland, stone, and perhaps a little water. A monument is just another pile of rocks. The true meaning of such things resides in our minds only. There, and there alone, lies the idea that a "garden" is a space *separated* from nature, designed, cared for and tended to, and belonging to somebody. Likewise, the true meaning of a particular "Peter" lies in everything you have been told about him, plus whatever you can add from your own personal experience. No less, and no more.

Interestingly, proper names like "Peter" do not even have a general definition. There are things that count as representatives of the class of trees, gardens, monuments and even democracies (as in, "Russia is not a democracy"). But proper names like "Peter" do not have any properties that all Peters share. Being called Peter does not turn you into a "Peter."

The range of language-dependent products of our inner world is impressive. They are notions that can be gleaned neither from the physical world around us nor from electrochemical or hormonal signals originating somewhere inside our bodies. They can be formed and transferred only by means of human language. The inner world of animals, too, entertains some concepts that have no basis in either physiology or the physical world. Say some group animal happens to have a mutation in its neural network that renders it just a little more sociable than usual. It will be more popular, which will give it an advantage over the other members of its group—and, in turn, greater chances of spreading its genes, mutation included. This might well be behind the signs of morality and altruism that have been found in certain species.

But that is as far as it goes. Because animals have no language, they have no means of developing these rudimentary inner world notions any further, or of combining them with others into more complex structures. They stay locked inside the individual's brain, in their original form. What is more, animals cannot develop or acquire new concepts over the course of their lives. This is something we humans do by formulating new ideas, reflecting on them, and pooling our knowledge by means of teaching and discussion—none of which can happen without language.

* * *

John B. Watson, the man behind behaviorism, saw the behavior of man and beast as a straightforward response to stimuli from outside the body. Think of it like a game of billiards, where the balls respond to the impact of the cue and collisions with one another. For Watson, humans and animals were little more than dead matter fitted with an adjustable memory. By means of this memory, a creature could develop patterns of responses to external influences.

Although things turned out to be far more complex, Watson's idea was not altogether without merit. At least some of the behavior of organisms operates along the lines of Watsonian habit formation. *Aplysia* is the prime example, the sea slug that learned to retract its vulnerable parts when exposed to electric shocks. The humble sea slug even shows vestiges of classical operand conditioning. The famous dogs of Ivan Pavlov, the nineteenth-century Russian physiologist, started salivating at the sound of a bell they had come to associate with the imminent arrival of food. Likewise, *Aplysia* learned to consider a barely there current a cue to retract its syphon, having apparently learned from experience that this preceeded the real blow.

Sometimes even human behavior seems to be equally uncomplicated. Picture a Sumatran farmer, making his way through the jungle toward his rice paddy. Suddenly there is a rustling in the *alag-alang* to his right. The farmer stops in his tracks and listens carefully. Is that the sound of . . . growling? It could be a tiger! His brain stem reacts immediately, sending him running away from the danger as fast as his legs can go. This is a direct response to an external stimulus. But there is more going on here, for the reason for this response—the farmer's fear of tigers—is almost certainly innate, something Watson wrongly claimed did not exist.

In fact, the inner worlds of both human beings and animals at birth are not like Watson's tabula rasa at all. They are preprogrammed with innate concepts, strategies and machinery for prepping incoming stimuli and streamlining, filtering, and integrating information. The brain stem, cerebellum, limbic system, amygdala, pituitary gland, and hippocampus all belong to an array of highly specialized structures in the brain that take care of all our bodily functions, but also the workings of our emotions and the construction of new memories. There is a certain amount of physical variation between our individual bodies and brains, giving rise to the tendencies, traits, and talents that are unique to a person. But as long as behavior results from these built-in systems and factors alone, it is quite predictable and understandable. Animal behavior is mostly like that. It is precisely the predictability and reliability of their behavior that gave mankind the opportunity to domesticate a few species, tame a few more, and keep the most dangerous and ferocious at bay.

Once the inner world starts activating parts of its contents of its own accord, as happened in early humans, everything changes. This is the first sign of real thinking and creativity, exerting an invisible but real influence on our behavior. As a result, we became less predictable. Our inner world ceased to be a mere reflection of inherited biological data and stimuli from inside the body and beyond. Instead, it began to create its own reality.

Notes

1. "Real" is a qualitative notion, a judgement that has no place in the material world, just like "fake," "failed," "overpriced," and so on. We can talk about a new, budget BMW as a "glorified Ford Mondeo," but this does not necessarily entail that a "glorified Ford Mondeo" is a type of BMW or even a car. For example, a journalist could perfectly sensibly write, "This administration lacks both ambition and prestige. The PM sits at the wheel of a glorified Mondeo, not the sturdy Bentley he should be driving."
2. Mouth aperture—that is, how wide you open your mouth—is also irrelevant. Doctors always tell people to say "aaah" because they usually open their mouths wide when doing so. But you can produce the vowels in *car*, *race*, and *beer* equally well with clenched teeth—so next time, surprise your general practitioner. Mouth aperture is only important for forming rounded sounds in words like *one*, *do*, and *go*; the nasal sound *m* and the plosives *p* and *b*, which require the mouth to shut completely; and the fricatives *f* and *v*, which require the lower lip to touch the upper teeth. It is this relatively minor role of the shape of the lips and the aperture of the mouth which, contrary to popular belief, makes lip-reading extremely difficult.
3. The answers are no, no, and yes. To transport that many bricks, a fleet of at least four thousand of the largest trucks available would be needed. Eight would suffice to carry the sugar, whereas a shopping bag would easily accommodate the salt.

4. The names of numbers, too, are a purely linguistic matter. Each language has its own relatively small set of basic number terms and a set of rules for deriving unique names for all other numbers from them. English has some two dozen basic terms: zero, one, eleven, twelve, hundred, thousand, million, billion, and so on. It creates further names through devices like the suffix "-ty," as in forty, fifty, sixty, seventy, eighty, and ninety, and by combining names in different orders, as in "six-hundred-forty-five" for 645. The fact that the system is riddled with exceptions—two tens are "twenty" and three tens are "thirty," not "twoty" and "threety"—shows that the rules are linguistic in character. This contrasts with the mathematical system of number symbols, of which the decimal system has exactly ten (the digits 0 to 9) and the binary scale only two (0 and 1). Words for large numbers like "million" are handy abbreviations for cumbersome formulas like "ten times ten times ten times ten times ten times ten," formulas that in turn are abbreviations of huge amounts of counting.

4

Phantom Limbs and Figments of the Mind

"Last year I had a terrible accident on my uncle's farm," says the young woman from Boston. She is speaking on the phone with Vilayanur Ramachandran, an Indian neurologist based in California. "I lost my left leg below the knee and I've had a phantom limb ever since." Ramachandran pricks up his ears. He has been fascinated for years by phantom limbs: arms, legs and other body parts which have been amputated or were missing at birth, but which the owner's mind clearly seems to think are still there. And he is not about to be disappointed. "Something really strange happened to me after the amputation that did not make sense," the woman says in his book *Phantoms in the Brain*. "Every time I have sex, I experience these strange sensations in my phantom foot."

The very next day, Ramachandran's phone rings again. This time it is a man, an engineer from Arkansas who lost the lower half of one leg only two months before. He has decided to reach out after reading a newspaper article by the doctor on phantom limbs. He too reports that, whenever he has intercourse, he feels things in his missing foot. Ramachandran starts to explain that this is not unusual; phantom sensations are par for the course in amputees. But the man cuts him off. "All that's fine, doctor. But you still do not understand. You see, I actually experience my orgasm in my foot. And therefore, it's much bigger than it used to be because it's no longer just confined to my genitals." An optimist might conclude that there is a bright side to amputation after all. But living with a phantom day in and day out is far from easy.

There is an explanation for such out-of-the-ordinary orgasms. The brain has a dedicated strip of cortical tissue responsible for processing most of the sensations coming from elsewhere in the body, like pain, heat and cold. This also includes proprioceptive information, or the sensations that help determine what does and does not belong in our

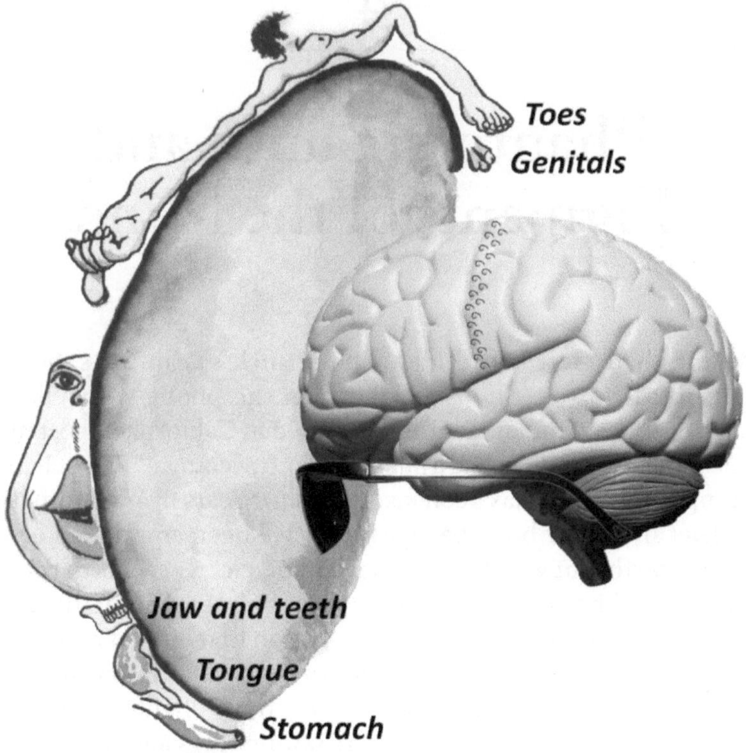

The somatosensory strip is indicated by a pattern of small black curls on the brain. On the left is a cross section of the left hemisphere along the somatosensory strip, with a rough layout of the areas dedicated to different parts of the body. Next to each other, buried deep in the cleft between the left and right hemispheres, are the toes and genitals.

body, how our body parts are positioned at any given time and how they move. Proprioception is what is temporarily absent when your leg is asleep: until it starts tingling, you feel as if the leg is not there at all. You can still move it, but the lack of feedback to your brain means you cannot feel what it is actually doing. This is why you cannot stand on it for fear of falling over: you have temporarily lost control of part of your body.

This somatosensory strip runs along the surface of the neocortex, starting deep inside the cleft between the left and right hemispheres and stretching all the way to the ears. Every body part that can receive and transfer the relevant signals has its own territory staked out on the strip. Those belonging to the left side of the body can be found on the right hemisphere, and those belonging to the right side on the left hemisphere. The size of this territory depends on the sensitivity of the body part. Body parts that are loaded with nerve endings, like the lips, tongue and fingertips, command relatively large areas on the strip. Relatively insensitive parts—our bottoms, say—are consigned to rather more cramped quarters.

Crucially, the areas on the somatosensory strip are not ordered in the same way as the body parts themselves. And so it is that, deep down between our brain hemispheres, right at the far end of the area representing our feet and toes, the territory of our genitals begins.

The fact that these areas are neighbors hints at why, in different cultures and down the ages, feet have been considered intensely erotic. What is more, it may explain the sensations reported by Ramachandran's callers. When a limb is amputated, the corresponding part of the somatosensory strip falls fallow. Neighboring areas, representing quite different body parts, can then encroach upon the unused space. This is why people who have lost a hand, for example, often report the sensation of having their missing fingers stroked when someone caresses their lips or cheeks. Like toes and genitalia, fingers and face are neighbors on the somatosensory strip.

All this is yet another illustration of the extent to which the world we experience depends on its representation inside our brain, even when that representation flatly contradicts the physical reality we can see with our own eyes. In the most literal sense, we shape our own bodies—and the world at large—in the image of our inner world.

* * *

Phantom limbs can cause their owners all sorts of trouble. One common problem is that amputees keep trying to use a limb that is no longer there. If you lose an arm, say, your reflexes and motor routines will simply carry on as though nothing has happened. So you will keep on trying to catch things and to pick up your coffee mug with your missing hand. Or worse, fend off a blow or break a fall. This behavior may wear off with time. But equally, it may well persist for years—or for life. Just imagine the toll in terms of bruises and broken crockery.

Phantom limbs can also cause intense pain. And in many cases, there is no way to treat or alleviate these phantom pains. Ramachandran describes patients driven to despair because their missing arm feels as though it is fixated in a plaster cast or a sling, and they cannot move it to release the tension or relieve the cramped limb. Others cannot sleep for the pain, or have trouble finding a comfortable position because their phantom arm always gets in the way. One patient was unable to prevent his phantom hand from constantly grabbing him by the throat. These are nightmares of the kind we like to think can only come out of Hollywood.

The cause of phantom pains is not always a mystery. When a limb is amputated, the nerves in it are severed. The frayed endings heal, but often little globules of scar tissue develop, called neuromas. These may cause all kinds of discomfort, from mild itching to excruciating pain. And there is no easy treatment. One option is to get rid of the neuromas by taking yet another slice off the amputated limb. This is a dubious solution at best, because there is every chance that new neuromas will simply form on the fresh nerve endings, not to mention the fact that the old neuromas may as phantoms continue to haunt their owner.

Neuromas are found in animals as well. They may form when an animal has its tail docked or its ears trimmed for aesthetic reasons, or as a side effect of trimming the beaks of bioindustrial fowl to reduce the damage caused by pecking. So chances are that these animals share the fate of human amputees. We cannot ask them directly, but we know for a fact that all mammals have somatosensory strips just like ours. What is more, a set of rather unsavory experiments on macaque monkeys showed the same kinds of cortical land grabbing seen in humans. Neurosurgeon Timothy Pons severed the nerves that sent signals from the macaques' forepaw to their brain. Years later, the hand area of the macaques' somatosensory strip showed activity whenever their cheeks or lips were stroked, suggesting that the hand territory had been taken over by the face areas. So far, so similar: these are clear parallels between the inner worlds of human beings and animals.

It is unclear whether this is also true of the so-called rubber hand illusion. This phenomenon is quite the opposite of a phantom limb. Instead of continuing to feel something that is no longer there, people experience an essentially foreign object as an integral part of their body. It is a simple trick. Picture someone seated at a table with one arm resting on the tabletop, the other out of sight below. Positioned directly above the subject's hidden hand is an ordinary rubber hand from a novelty shop. Now, if someone starts stroking both the real and the rubber hand at exactly the same time and in exactly the same way, the subject soon begins to regard the rubber hand as her own, so much so that she will jump a mile when a hammer comes crashing down on it.

Some illusions go even further. In 2008, researchers in Sweden discovered that it was not hard to make people believe they were inhabiting an entirely different body. They kitted out a mannequin with small cameras at eye level, and the subjects wore glasses with little screens instead of normal lenses. As a result, they effectively saw the world through the "eyes" of the dummy. When the subject and the dummy were stroked at the same time, in the same place, it did not take long for a strange kind of transmigration to occur. The strength of the illusion became apparent when the dummy was replaced by a real person with cameras stuck to his brow. When that person shook hands with the subject—who saw this as his own arm shaking hands with "himself"—he still could not escape the feeling of being "in" the other person's body. His inner world seemed to have temporarily taken up residence in a different body.

Whether animals might be susceptible to similar illusions is an open question. All we can say is that animals react to visual stimuli in totally different ways from humans. Dogs and cats, for instance, do not even recognize themselves in the mirror. So whether they could "recognize" a rubber paw as their own is a murky question indeed.

* * *

One of Ramachandran's most impressive cases was that of a girl born without forearms or hands. What she did have were two phantom hands, which—in her mind, at least—were positioned about six inches beyond the stumps of her arms. She even used the fingers of her invisible hands to count on. Since she was born with her handicap, these phantoms cannot have arisen from memory or experience. It was just her inner world functioning completely independently of the physical facts.

So in some cases, the fact that our bodies do not always conform to our inner world's ideas about them can be useful. But that is not always true. In another case, a woman kept falling out of bed, over and over again, every single night. To get to the bottom of this, she was eventually put under constant surveillance. It turned out that as she slept, she would tug violently on one of her arms until she succeeded in flinging it out of her bed—and the rest of her with it. When asked why she did this, she replied that the arm was not hers. It was an intruder into her bed, an interloper that had to be removed. Out with it!

For most owners of phantom limbs, the source of their troubles is a purely physical trauma. This woman, by contrast, had serious mental problems. Although her body was perfectly normal, her mind did not recognize parts of it as her own. Her inner world had derailed in such a way that it harmed her ability to deal with the world, including her own body. Bizarre as it may seem, such cases are not nearly as rare as you might think.

Inner worlds frequently go out of whack as a result of brain damage. For young people, the most common cause of severe brain damage is crashing a motorbike with no helmet on. Among older people, as our life expectancy increases, brain infarctions are becoming more and more frequent. The resulting damage is altogether unpredictable: no two cases are exactly alike. Parts of the body may be paralyzed. The patient may suffer from personality changes, something their family and friends have an especially hard time swallowing. And if the left hemisphere is afflicted, the patient may develop aphasia. This can impair the ability to speak, so that patients have difficulty forming and pronouncing words and sentences. Some are left with slow, hesitant, or stilted speech. Others can no longer produce anything remotely resembling a real sentence. Still more are so badly afflicted that they can only produce one or two words, or even a single sound. But even then, they may still be able to hear and comprehend perfectly what others are saying. In other cases, aphasia impairs not the ability to speak but to understand language. Patients may still chatter away like the best of them, but not understand a word of what is being said back.

The kind of brain damage most relevant here is an affliction called neglect. Neglect occurs if your brain has been damaged in such a way that you no longer notice part of your surroundings. This is not like being blind or deaf, where you are well aware that you cannot see or hear. If you are suffering from neglect, you are completely unaware that anything is amiss. As far as you are concerned, there is nothing wrong with your powers of observation.

It is important not to confuse neglect with denial. People who are in denial cling to a self-image that clearly contradicts reality. In one such case, a woman who had lost an arm through amputation doggedly maintained there was nothing wrong with her. Why, she could even clap her hands! Whenever she was challenged to do so, she would find some subterfuge for why she could not at that particular moment. This is a classic case of denial. Confronted with the problem they just cannot come to terms with, patients will go to any lengths, devise any pretext, to avoid showing themselves up to themselves. This does not mean they are lying; it is just that they are as convinced of their self-image, however warped, as the rest of us.

In neglect, there is no subterfuge. Why should there be, if there is no problem to begin with? The best-known examples are people who have lost one half of their field of vision. Ask them to draw the face of a clock, and they will happily put pen to paper, but produce only the right half of the clock—or the left, as the case may be. Yet they will be convinced they have come up with a faithful representation. Put them in a room and ask them to sketch what they see, and they will draw only half of the room. Put them at the other end of the room, and they will draw the other half. In neither case will they have the faintest suspicion that anything is wrong.

Cases of neglect are not always as easy to spot as this. People may lose just a tiny fragment of their visual field, or a few elements of spatial orientation. This can give rise to difficulties with activities like reading, writing, watching television, and generally getting about that are both irritating and seemingly inexplicable. Neglect comes in all kinds of different guises, all equally sneaky. What they share is that parts of the person's inner world have been damaged beyond repair. Yet they are not aware of this, nor can they be made aware of it. The inner world operates independently of the world outside, but not independently of itself. From our inner world, we can observe ourselves just as easily as we observe other people. But we cannot observe our own inner world itself. We *are* our inner world.

* * *

Dr. Ramachandran's consultation room is a treasure trove of phantom phenomena. Among them is a man who has come in to complain about the nails on his phantom hand. It has been seven months since the amputation, and the hand is driving him up the wall. It cramps up, he explains to Ramachandran, which causes his phantom nails to dig into his phantom palm, causing very real pain. If phantom hands were capable of bleeding, his would certainly have done so. Again, it seems too bizarre to be true, but this and similar complaints actually are not all that unusual among hand amputees. Even Admiral Horatio Nelson, the British hero of the Napoleonic wars who lost his right arm at Tenerife in 1797, suffered from something like this. He considered his phantom pains "proof of the existence of the soul." As we shall see, the one-armed aristocrat was not far wrong.

Ramachandran came up with a plausible explanation. When you clench your fist, the effort you put in is tempered by the nerve signals that are sent back from the hand to the brain. Nothing out of the ordinary happens until your fingertips reach the palm, but when your nails start digging into your own flesh, the nerves inside the palm start sending warning signals up to the brain: enough, ease the pressure, that hurts! But if the hand in question is a phantom hand, there are no warning signals. So the brain just keeps on sending orders down to the hand to contract—harder, harder, harder!—and nothing tells it to stop.

The question is, where does the pain come from? After all, there are no real nails, no palm, no hand at all. Perhaps, Ramachandran proposed, it was the endless stream of contraction orders itself that was experienced as pain. But this does not explain what causes the particular sensation of nails digging into flesh. Perhaps, instead, a permanent feedback loop had formed through the many times the man had made a fist during his life before losing his hand. The brain's instructions to clench the fingers might then have become automatically linked to signals sent back by the palm as it begins to feel the pressure of the fingernails. This could be the same mechanism that means we do not feel pain when we imagine clenching our fingers, even when we picture doing it so hard that it hurts. The normal signals from our hands tell our brain there is nothing wrong. It is this flow of reassuring signals, according to Ramachandran, that is absent in amputees. As a result, the "memory" of pain is experienced as real pain.

Ramachandran offers a similar explanation for Charles Bonnet syndrome. Bonnet was an eighteenth-century Swiss naturalist who described the lively hallucinations that haunted his grandfather. The old man suffered from failing eyesight, but was otherwise in good shape. Hallucinations are not unusual among old people whose vision is deteriorating. Mentally there is nothing wrong with them—and yet they see things that just are not there. Sometimes it is nothing more than abstract shapes. But others see pets, old acquaintances, or even Disney characters. The images tend to be razor sharp, even when the person's eyesight is no longer what it once was. In fact, this unlikely clarity is often how people first realize they must be hallucinating.

Ramachandran's explanation is that when our eyesight starts failing, our visual system becomes underemployed. To keep itself busy, it starts generating its own impulses, drawing on everything it can find in the huge archives of our memory. And that is how a perfectly sane person suddenly sees Bambi trotting across the shoulders of their dinner companion, or dozens of tiny, brightly colored wheelbarrows strewn across the floor.[1]

* * *

Each of these cases shows, in its own way, the independence of our inner world from the world of flesh and blood and steel and stone around us. They show just how smoothly the inner world handles all kinds of ideas, concepts, objects, and categories that do not—and sometimes cannot—exist in the physical world, up to and including missing body parts. At the same time, they illustrate how healthy people can keep in touch with reality in spite of the weak link between this reality and our inner world.

As a rule, inner worlds become unhinged for one of two reasons. The first is that parts of a person's inner world have been destroyed, as in the case of neglect. The consequences in such cases are grave and permanent, but also limited and linked directly to the damage to the brain. The other reason is a lack of feedback from outside the inner world, that is, a lack of sensory input about what is going on both inside and beyond our body. This is how phantom problems come about, as well as the relatively innocent hallucinations of Charles Bonnet's syndrome. As we shall see, a lack of feedback can also cause entirely different phenomena, and there are other reasons why inner worlds go haywire. For now, the important point is that we need feedback to keep our inner world in sync with external reality.

What we know about neglect also helps us answer the classic conundrum that cognitive scientist Daniel Dennet disdainfully calls the Cartesian theater, also known as the homunculus problem. The sensation of watching ourselves through our "mind's eye" lies at the very core of our self-awareness, conscious observation and reasoning. We "see" ourselves contemplating, weighing options, and acting on our conclusions just as we observe an actor in a film or follow a character in a book. It is as though there is a homunculus, a little man—or woman, as the case may be—sitting in a theater inside your head. He watches you act out the story of your own life, interprets it, and occasionally even tries to rewrite the script. This is an idea we all share. And yet it cannot be true, because to be able to do its work, the homunculus would have to have powers of observation of his own, not to mention the capacity for conscious thought and self-awareness. But that would entail the existence of a homunculus inside the homunculus, whose even tinier head would accommodate a third homunculus, and so on ad inifinitum. That is not only absurd; it is also impossible. And yet we cannot shake the sensation of a mind's eye watching every move we make. Neglect gives us a way of reconciling these two notions.

Let us first acknowledge that we really do watch ourselves. It is not an illusion or some kind of metaphor. It is our inner world, our personality, our "self," or whatever you prefer to call it, keeping track of how our body interacts with the world outside it. It has somehow uncoupled itself from external reality and become autonomous, which is what enables us to look at ourselves as someone else. We have all become a permanent fixture of our own inner world, actors and actresses on the stage of our internal mental universe. The inner world can also see its own linguistic products: the conscious lines of reasoning, arguments and interior monologues that help us get through the tasks of each day, whether we are vacuuming the house or climbing Mount Everest. This explains our conviction that conscious thought involves words: that kind of reasoning is indeed a function of language.

For all its autonomy, there are still things our inner world is not capable of. Most importantly, the phenomenon of neglect proves that it has no self-awareness. When parts of it are damaged or destroyed, it never finds out. Nor does it ever discover that all those interior monologues and arguments milling around in it are its own creations. This is what prevents an infinite regression of mental theaters populated by ever-smaller homunculi. Quite unaware, our self-reliant inner world is both the source and the essence of our self-awareness. Consciousness is the product of cooperation between the inner world and the body—and so consciousness is about the body, not about the inner world itself. This we know from the rubber hand illusion and, especially poignantly, the Swedish transmigration illusion. If you are deluded into thinking you are inhabiting a different body, even an obviously fake one, you soon begin to recognize that body as "yourself."

So both our sense of self—the "I," so to speak—and our consciousness turn out to be the fantastic fruits of an inner world that has broken free of reality and acquired a life of its own. It is this extremely tenuous connection between the inner world and external reality that is our most profoundly human trait.

There is a danger here, however. For our bodies to stay in the fight for their continued existence, we must retain at least *some* correspondence between the figments of our inner world and the facts of the real world. It is a perilous equilibrium, maintained by means of sensory feedback, which can easily be disturbed or fly off the rails altogether. If worse comes to worst and our inner world loses track of reality entirely, we can only survive by taking refuge in a carefully structured, man-made environment.

* * *

The growth and emancipation of the inner world marked the beginning of typically human kinds of intelligence, of creativity, and of free will.

Intelligent behavior takes into account more than just the information that happens to reach the brain at a particular moment in time. It is behavior that takes advantage of memories of prior experiences and other available knowledge. The broader the scope, the more intelligent the behavior. Creativity is a special case of this. We speak of creativity when existing elements in the inventory of the inner world are combined in novel ways, producing a new, meaningful result. Creativity is the way the inner world extends its inventory without drawing on the world outside.

Two factors determine just how intelligent and creative any species can be, humans included. The first is the size, scope, and quality of the materials available for thinking. The second are the tools for manipulating and combining these materials. In both respects, the most intelligent of animals are still severely limited compared to human beings. Even those species capable of acquiring inner world products cannot do much with them, because they lack the necessary tools for structuring and analyzing them. But humans do have such a toolbox. It is called language.

Language does more than just enable us to develop inner world products. It also gives us the fantastic ability to construct highly complex arguments, ideas, and trains of thought. Beyond that, from language sprang logic and mathematics, the keys that opened the door to technology, culture, and civilization, and—for better or for worse—enabled mankind to become masters of the world. Like any language, logic and mathematics are systems for manipulating symbols—every one of them inner world products. And just like English, Urdu, or Inuit, they are instantiations of the innate skeleton called universal grammar that we all share. What makes them special among languages is that they have been stripped to the bone so vigorously that all that remains is an extremely limited, rigid grammar, a vestigial, jealously guarded vocabulary, and none of the flexibility and ambiguity of the languages we ordinary mortals use to chat among ourselves. Whatever this willful austerity takes away, it makes up for in precision.

One of the great gifts of language, then, is that it has enabled us to become vastly more intelligent than all our animal relatives. The other great gift—or curse, if you prefer—is free will.

Our inner world gives us the greatest illusion of all: the unshakeable conviction that we exist independently of everything else. It makes us feel separate from the outside world, and even, to some extent, from our own bodies. Hand in hand with this goes the conviction that we make our decisions of our own accord, on the basis of conscious, rational deliberation.

Deep in our hearts, of course, we know this is not entirely true. Say you have to buy a new refrigerator. This is no small matter. Fridges are like family members; they stick around for years. So you spend weeks plowing through brochures, websites, and consumer reports, and having endless conversations with store clerks. You boil it down to three final candidates, all essentially the same, each ever so slightly different. But inevitably, the final choice is random. Either you jab an exasperated finger at one and say "All right, that'll do." Or you turn to your partner and say, "Darling, you choose." It is impossible to say what the deciding factor is, or whether the whole procedure is really any different from just rolling dice.

Or think of the pub, an institution you normally visit of your own free will. The bartender invariably expects you to order a drink. Usually you ask for your favorite tipple, but today, on a whim, you go for something completely different. What exactly underlies your decision remains a mystery. Is it an expression of the unbounded freedom of random choice? Is it the subconscious pressure of assorted hormonal and other biochemical factors? Is it a matter of rational, but unconscious reasoning? And if so, how "free" was your assessment of the factors involved?

This is not getting us any closer to finding out what free will really means. So let us look at it from another angle. The "will" part is easy: "will" is the urge to do whatever emerges from the evaluation of all factors pertinent to a decision. It is shorthand for both a conscious decision and knowingly acting upon an urge.

The meaning of "free", however, is a thornier issue. It can hardly refer to the boundless, and thus essentially inconsequential, freedom of random choice. Decisions taken with no underlying rationale whatsoever are not merely stupid. They are meaningless. Moreover, they contradict our own ideas and feelings about how we go about exercising our free will. Instead, we can think of an act of free will as a considered choice on the basis of our own experience, wishes, and interests, without undue pressure from external factors. We do not mind other parties having an influence on our decisions, as long as it is not excessive.

Many people consider modern science to be a threat to the idea of free will. Science, they say, reduces us to automatons whose behavior is completely determined by innate, hereditary traits. They worry that as scientists probe deeper and deeper into the biological nooks and crannies of organisms, more and more aspects of human behavior are attributed merely to genes, hormones, and vitamins. This is a misguided, irrational fear. Even if we did prove to be genetically controlled robots, we cannot blame science for finding that out, any more than we can blame it for the color of moonlight.

That said, there is no denying that science sometimes unearths quite disconcerting patterns of human behavior. Psychologists and economists have shown that we tend to interact with each other and our environment in highly stereotyped ways. We do this even when we think we are not, and in ways that are hardly in keeping with our ideas about the sovereignty of our emotions. For example, experiments have shown that when it comes down to it, we do not really care how much we earn, as long as we make more than our neighbors. That is hardly how we like to view ourselves, but apparently our inner world entertains certain opinions about us as well.

Neurologists, too, have cast doubt on the plausibility of free will. Picture yourself faced with a choice between two alternative actions. Seconds before you even become aware that there is a choice to be made, a particular motor activity in your brain starts gearing up to help you make that choice. This gives rise to the sneaking suspicion that we lack free will altogether. Instead, our opinions are served up to our consciousness ready-made, from deeper, more primitive systems that are much quicker to respond to sensory input. This would make our consciousness something like a ceremonial king. The real decisions are made over his head. All he gets to do is add a symbolic signature amid a lot of pomp and ceremony.

Do the discoveries and conjectures of these scientists actually imply that we do not really have free will? On the contrary. Every single one of us has an inner world that differs from everyone else's, both in its hereditary endowment and in the experiences and memories it has acquired. There will be similarities, particularly with the inner worlds of our close relatives, friends, colleagues, and compatriots. But no two are identical. Myriad details make any given inner world differ from all others in unpredictable ways. They are dynamic, and they operate without regard to the laws of the physical world, which means that elements contained within them may gain or lose prominence and assume different roles in our musings, on grounds that are ultimately of their own choosing. And this is what makes each of our decisions, every expression of our will, into something unique. Something that sprouted from us could not have sprouted from anybody else. It is utterly our own.

That is freedom. Of course, we usually make sensible, efficient choices. And, of course, that makes us somewhat predictable. But even if our preferences are generally similar to everyone else's, this does not prevent us from arriving at completely different conclusions from everybody else every now and again, for reasons buried in the deepest recesses of our own unique inner world. Most drivers will wait for a green light at an intersection, even if there are no other cars in sight. But once in a while, you may decide to jump a red light, according to your assessment of the actual situation—say, you are running late for a date with someone you just fell head over heels in love with. Or think of two people sentenced to the same punishment by the same judge on the same grounds. One decides to appeal, while the other decides to take it on the chin and be done with it. We each make our own decisions, based on our own feelings and considerations.

This is what machines are incapable of: making decisions based not simply on the sum of predictable factors of some predetermined relative importance, but rather on the demands of the current situation. You weigh considerations that are important to you, regardless of whether they derive from your genetic endowment or not, and regardless of whether you are even aware of their influence. *That* is free will.

* * *

If free will *results* from the combination of the inventory of our inner world, intelligence, creativity, and the structuring powers of language, it cannot have given rise to any of these. Each of the faculties involved must have developed independently, for its own good reasons. And every step in their development must have been justified by an immediate evolutionary advantage. After all, evolution does not plan. So how did it all start?

Obviously, everything must have begun with a memory and a talent for learning. These are traits shared by even our modest *Aplysia*, which testifies to their evolutionary benefits even in the simple world of the sea slug. And it is not hard to see why.

Imagine a very primitive animal, many hundreds of millions of years ago, with a very simple nervous system of only a few hundred neurons. These neurons react to all kinds of impulses by firing and then going back to sleep again. That is, from time to time a signal travels through the network, causes some physical reflex or other, and then peters out.

Now suppose that for some reason, one such signal pattern is not extinguished entirely, but has a sort of afterglow. It seems reasonable that if a similar impulse were then received, this lingering signal pattern would be activated again much faster than it would take to produce a new one from scratch. Or it could be reignited by a much weaker impulse. And if the effects of this process were beneficial, it would enhance the animal's chances of survival.

This, very broadly speaking, is how we think all biological memories work. External influences give rise to signal patterns in a network of neurons, which take time to fade away after the stimulus has passed. If a similar stimulus comes along during this "afterlife," one of two things may happen. Either the lingering signal pattern is rekindled, and so the animal's response is faster, or a second, largely identical pattern comes into existence, making the animal's response stronger.

Naturally, being able to respond to stimuli more quickly and powerfully is an advantage. So it is not surprising that creatures from all corners of the animal kingdom have developed memories. There must have been serious selectional pressure favoring those individuals with the longest afterglow period. Eventually these patterns would come to be retained indefinitely, at which point they ceased being mere patterns, and became memories.

Memory allows our behavior to become more flexible and subtle, in other words, more intelligent. In a well-stocked memory, a stimulus will reactivate many similar circuits, which together determine the response. As such an inner world becomes ever larger and more complex. It increasingly draws only on itself in generating behavioral responses. Accordingly, its ties to its environment loosen.

In us humans, our inner world has grown to a truly prodigious size and reached a level of complexity to match. It took quite a lot to make all this possible. But none of this would have happened if our hominid predecessors had not first developed a very, very large brain.

* * *

Two to three million years ago, the bodies of our forebears began to change in a way that would put them on a unique course among animals. Little by little, over thousands of generations, the heads and brains of these bipedal inhabitants of the African savannah grew larger and larger. Lucy, an *Australopithecus afarensis* who lived around the time this development began, still had to make do with a brain that weighed little more than a pound, much like that of a modern chimpanzee. One and a half million years later, Lucy's descendants had morphed into *Homo erectus,* sporting brains roughly twice the size of hers. Our brains today, yours and mine, weigh around three pounds. Those of Neanderthals were a little bigger still.

What caused this spectacular development remains a mystery. At least as many propositions have been put forward as there have been rows and controversies over the matter.

One problem is that, from an evolutionary perspective, overly large brains and the equally large skulls needed to house them are fairly improbable. For one thing, these bigger heads saddled the women of the species with serious labor pains and a high risk of death during childbirth. Despite all of Mother Nature's trickery involving fontanels and shifting skull bones, there is a limit to what the birth channel can take. As a result, humans started being born before they were ready for the world, leaving our newborns unable to fend for themselves for far longer than any other species. And yet those brains just kept on growing.

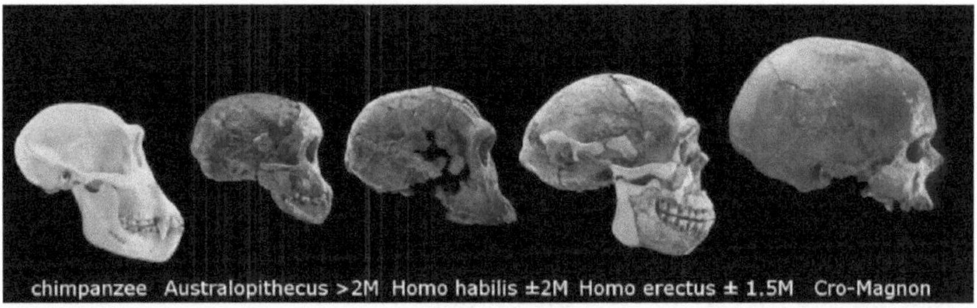

The leftmost image is the skull of a modern chimpanzee, roughly equal in size to that of our common ancestor. To its right, in order of seniority, are our own most prominent forefathers.

Why? What made having such an enormous head worthwhile, in spite of all the risks for mother and child? Certainly not the intellectual achievements of the brains inside those heads. For eons after our forebears' heads began to grow, very little was achieved besides a few small innovations in making stone tools. They were only tiny bouts of creativity between which nothing changed for hundreds of thousands of years. This is a leisurely pace that would barely challenge modern apes, and they are not burdened by excessive dollops of grey matter.

The only explanation that seems to hold any water is one offered by American anthropologist Dean Falk, and even that account is by no means generally accepted. The bones making up our skulls have been riddled with foramina, minuscule openings, since the growth trajectory started. Passing through these foramina are tiny blood vessels. Falk proposed that this network of vessels carrying blood from the brain to the skin outside the skull and back functions as a radiator, helping control the temperature of the brain. Temperature is crucially important when it comes to brain tissue: without some dedicated cooling device, brains any larger than Lucy's would overheat. So, Falk concluded, it was these foramina that opened up the path toward larger brains.

At best, however, Falk's proposal is only half the story. It takes care of one problem—how it became physically possible for our brains to grow so large—but it does not do away with the many other objections against massive skulls, like birthing difficulties, or the very important fact that brains gobble up more energy than gas-guzzling Hummers. Our brains account for only 2 percent of our body mass, but even at rest they burn more than 20 percent of our body's total energy consumption.

What is more, the human head is responsible for considerable heat loss. Dozens of experiments have shown that, even at room temperature, no less than a third of all heat that escapes our bodies (except through sweating) is lost via the head. In cold circumstances, this can easily increase to around half of our total heat expenditure. So it seems reasonable to assume that the primary function of the head is cooling. In a twisted way, this vindicates Aristotle, who conjectured that the brain existed merely to cool the blood.

Cooling therefore seems to be a promising line of investigation if we want to identify the evolutionary advantage of a larger head. Cooling capacity is a gradual thing, and in a hot environment, even the slightest improvement of a cooling system offers immediate benefits. And hot it certainly was in the African savannah of three million years ago, where the sun beat down mercilessly, day after scorching day.

It was an environment well suited to diurnal bipeds, creatures that walk on two legs and go about their business during the daytime. Their upright posture gave them the best possible view of the landscape. It also reduced the amount of skin directly exposed to the sun, from their entire back to only the top of their head and shoulders. Since the rest of their bodies only caught the sunlight at an angle, they could shed most of their fur in the battle against overheating. Only the tops of their heads remained covered with hair, just as it is today.[2]

With their heads taking the brunt of this solar onslaught, forced cooling of both the cranial surface and the contents of the skull was necessary. The denser the network of foramina, the better. A larger head was conducive to this end, for it meant a larger cooling surface. But its larger shadow also provided somewhat better protection for the body beneath it. In areas where the sun is high up in the sky, a good head of hair works like a natural parasol.

So this gradual growth of the hominid head must at first have had a markedly positive effect. Improved temperature regulation more than warranted an increase in the energy consumption of the enlarged brain. Difficulties in childbirth were still a thing of the future. But with time, that would change. So something had to happen to ensure the continued advantage of ever-larger brains; otherwise, the process would simply grind to a halt. That something was the emergence of new and more powerful mental faculties.

The crucial point here is that when the brain started growing, not all parts of it increased in size. The older systems hardly changed at all. But the cerebrum, which is like a sponge covering the older limbic system, cerebellum, and brain stem, became larger and larger. Its creased, pinkish brown surface is what we call the cortex or grey matter, because it turns grey when preserved after death. It consists mainly of the cell bodies of neurons. Below that is the white matter, with its wild tangle of runners sprouting from the neurons in the cortex. These runners, or axons, connect one neuron to another and pass signals between them.

We can largely thank this white matter for our enlarged brains. It was not so much the number of neurons that multiplied as the number of connections between them. With so many more connections, the brain could accommodate many more, and more complex, signal patterns. All those new connections provided a wealth of new pathways for neurons to fire and excite other neurons. It was as though an immensely intricate Internet were gradually awakening and turning a bunch of isolated, sleepy offices spread around the cortex into a thriving industry.

This slow revolution paved the way for many new forms of cooperation and activities hitherto unheard of. These developments justified the drawbacks of increased brain size in terms of energy consumption and the mechanics of childbirth. And later, one of these developments would prove crucial to the genesis of human language. It was the ability to throw.

* * *

Finding food in the dry African savannah where our bipedal predecessors lived was no easy task. Certainly it was not as simple as it had been in the wet, lush rain forests their forebears had abandoned. Those primordial tree dwellers probably ate fruit and whatever else they came across among the leaves and branches, including the occasional small animal. Their way of life looked very much like that of modern chimpanzees. The open savannah was not so generous, but the relative dearth of food was compensated for by a larger territory to forage in. These early hominids also adapted their diets: tubers and root vegetables replaced fruit as their staple foods, with a bit of meat thrown in now and then.

These were momentous changes in their circumstances and way of life. Two things made these changes viable. First, they still had the nimble hands of their tree-dwelling forebears. And crucially, they developed the new art of walking. This freed up those clever hands for gathering and carrying food, and hunting small animals like mice and rabbits.

Catching a rabbit barehanded, it is fair to say, is no small feat. Rabbits are fast and nimble, traits they share with most other animals that are both edible and not too dangerous. Our hunter-gatherers were relatively slow in comparison, and clawless and fangless to boot. So rabbit meat must have been a rare delight on a predominantly vegetarian menu. Compared to meat, vegetables are low in energy and take a lot of time and work to digest. This would not have made life any easier for early hominids.

And yet ultimately, they were remarkably successful. Around two million years ago, *Homo habilis,* whose brain had increased to about one and a half pounds, had spread out across the whole of the African continent. Half a million years after that, we find *Homo erectus,* a yet more advanced hominid, all over not just Africa but also Europe and Asia. Endowed with just over two pounds of brain matter, they inhabited dry regions and wet, hot areas as well as cold. In short, they lived wherever they could survive without clothing or heated shelter. Interestingly, they also turned up in environments that cannot possibly have yielded enough roots, tubers, nuts, and seeds to keep alive.

There is another source of sustenance, however, which is ubiquitous: the meat of small mammals and birds. Everywhere, from the chilly tundra to the scorched sands of the great deserts, edible animals can be found that have adapted to even the most inhospitable circumstances. But to take advantage of this, our amateur hunter had to become a lot more competent.

That we still crave a juicy steak or a crispy rasher of bacon testifies to the fact that *Homo erectus* did become skillful enough to avail himself of a meaty diet. In itself, this is not very surprising. The really remarkable thing is that it happened long before the invention of hunting gear like bows and arrows. So how did these creatures, whose speed and agility posed no serious threat to their prey, manage to catch enough of them to sustain themselves?

The best answer we have to this question is one put forward by the American neurobiologist William Calvin. Take today's great apes. They are still bound to their original tropical habitat and diet, whereas humans have shaken off the shackles of a single environment and a uniform diet. According to Calvin, the difference may well lie in our different hunting and fighting tactics, specifically, in tactics that are available to us, but not to apes.

The most obvious difference is our endless variety of lethal weaponry. But that cannot be the smoking gun, so to speak, because our ancestors were happily hunting away before any of those tools were invented. Equally obvious is the fact that we have language, whereas apes do not. Again, this does not get us anywhere. Language is without a doubt the most powerful tool at our disposal, but it is not especially useful for hunting. It is clear from the sophisticated, organized hunting of wolf packs, for instance, that language is not a prerequisite for effective cooperation. Wolves can join forces to isolate, hound, and wear down their prey, all without a word. And what do human hunters talk about when they go out to bag an animal or two? Ordinary life, just as they might do in the pub or around the coffee machine at work: relationships, jobs, money. Once the hunt is on for real, hunters keep mum. Prey may not understand human language, but they can certainly hear it and flee.

Calvin pinpointed a third difference between humans and apes, which does have enormous significance. We can throw, and not just throw, but throw accurately. If you have ever visited a zoo, you will have seen apes tossing twigs and stones in the air and throwing dung around. But they do so more or less at random. We, on the other hand, can take aim with a projectile, use our arm as a kind of sling, and hit a distant target with great force. By acquiring the ability to do so, our fangless, clawless predecessor suddenly had at his disposal a deadly talon that could reach twenty yards or more.

Even a slight increase in throwing accuracy greatly improves the chances of hunting success, with all the evolutionary advantages this entails. On the other hand, it requires an impressive amount of rapid and complex calculation. Within fractions of a second, you have to conceive and execute a complete operational plan involving dozens of muscles, taking into account the observed course and speed of the prey, the estimated weight of the stone in your hand, your own position, direction, and speed, and so on. The margin of error is tiny.

Actually, speed and pinpoint accuracy are not among the strengths of our neurons. By nature, neurons are somewhat sluggish and sloppy. But the brain can increase its working speed and accuracy by making large numbers of neurons do the same work. The combined effort of hundreds of neurons, each individually quite inaccurate, can result in a surprisingly precise signal. And having several circuits work in parallel is a sure-fire way to ramp speed up a notch.

In *Homo*, the exorbitant brain capacity needed to do all this happened to be available, thanks to their larger brains. The gains of the new throwing faculty also easily offset any increase in energy consumption: a better thrower enjoyed considerably more extra calories and proteins than were needed to sustain his hardworking brain.

This was also the point when, probably for the first time, another remarkable phenomenon came about: lateralization. Our throwing ability is far more highly developed in one brain hemisphere than in the other. Nine people out of ten throw accurately with their right hand only, like so many other things that require subtle control and great accuracy. This, too, must be a direct consequence of Darwinian selection, and it is not difficult to see why. Being able to throw accurately with both hands adds little to your chances of hitting a rabbit, but it does require almost double the amount of brain tissue. So in effect, individuals with only one fully developed "ballistic computer" were in fact better off than their ambidextrous peers. This is probably where the asymmetry of hemispheres began that is so characteristic of the modern human brain.

All right-handers—so 90 percent of the population—have their ballistic computer in their left brain hemisphere, because the left hemisphere controls the right side of the body. That same hemisphere also houses the super-fast calculation programs of the language faculty, concentrated in and around Broca's area, just in front of and slightly above the ear. And it is the locus of arithmetic and other calculation tasks. All this suggests that the ability to throw was not only a huge step forward but also lay the foundation for many more typically human capabilities that would not develop until much later.

* * *

Throwing was not the only skill made possible by larger brains. Roughly 1.7 million years ago, the first real tools appeared: stone scrapers, cutters, and beaters. The tools the hominids made were not just more sophisticated and of better quality than those of modern apes and other animals. Whenever an improvement came along, it would spread more or less wherever hominids could be found. That is something we do not see in apes. They do sometimes invent new or better tools, or show other behaviors we might call cultural innovations, but these typically remain confined to the immediate group. Certain troops of West African chimpanzees have come up with better techniques for cracking tough nuts with stones, for instance, but neighboring groups show little interest in copying them. Perhaps the best example of this cultural inertness comes from the famous macaques of Koshima Island in Japan. Toward the end of the 1950s, some of the macaques were seen washing their sandy sweet potatoes in fresh water before eating them. Their young followed suit, as in turn did their offspring. But the behavior never really spread to other groups on the island, much less beyond. Today, it even seems to be on the wane, as if the macaques are gradually losing interest.

Singing and sensitivity to rhythm are two more capabilities that may have emerged quite early thanks to the calculatory prowess we developed for throwing. These can be found in other species as well—birds, for instance—but not so much among our closest relatives. For humans, however, song has been a potent means of strengthening group cohesion for as long as we can remember. The same goes for rhythm. Think of the effects of chanting slogans at a demonstration or bellowing out the club song at your local football stadium. Think of the compelling power of war drums, the captivating ritual of tribal dancing, or the enthralling pumping of pop music. Top DJs earn their crust by providing a bath of rhythmic noise for thousands to lose themselves in, with or without the help of a pill or two. "Dance" is rhythm, music divested of its melodic and harmonic trimmings—and it is immensely popular. In 2008, a British court banned a performance by the rock band Babyshambles for fear that their "dangerous" rhythms would whip the crowd into a frenzy. Clearly, this was an overreaction. But it was one based on sheer awe for the power of music and rhythm.

In any event, something big was going on in the course of those two to three million years when hominids' brains just kept on growing. Unbeknown to them, their brains were developing all the skills and capabilities that would someday combine to produce the language faculty we have today. In and of themselves, these advances brought them unprecedented success and resilience. So how far might they have come without language? And what triggered that final, giant leap that ultimately turned them into us?

Notes

1. Something similar seems to happen to people who develop hearing disorders in old age: they hear sounds and voices that are not actually there. Such auditory hallucinations may be even more unsettling than visual ones. It is clear that something is wrong if you wake up one Sunday morning and find Donald Duck waving at you from the tree outside your bedroom window—as happened regularly to my uncle. But if you do not hear all that well anyway, how can you be sure whether a sound you hear is real or not? And how can you tell others you are hearing voices without them thinking you have become dangerously unhinged?

2. And this vestigial fur is only poorly attached, as men are all too aware. Our most permanent fur is in the crotch and armpits, and on men's cheeks and chins. It has a different structure from the hair on our head, probably serving sexual ends such as gathering and retaining odors. Incidentally, it is precisely the armpits and crotch where many animals, like dogs, have little to no hair. These animals have no eccrine glands, which humans, primates, and horses have to regulate their body temperature by producing sweat. They have apocrine glands instead, which produce less moisture and more scent.

5

Shamans in the Shadows

The world of Tuc d'Audoubert is not for the faint-hearted. From the lush, overgrown entrance to this cavern in the French Pyrenees flows a small stream. It is the sort of magical place where elves or fairies might live. But those who venture inside embark on a creepy-crawly journey to the innards of the earth, first by water, then through passageways, down shafts, and up chimneys. At the very far end, almost half a mile into the mountain, a remarkable surprise awaits: two bison, each about the size of a dog, molded from clay, waiting patiently in eternal darkness. They are not freestanding, but jut out from a large stone backdrop, like thick reliefs, as fresh as the day their maker's wet, gentle fingers eased away the last imperfections. There are footprints in the damp clay floor of the cave around the effigies. Some must have been made by adults, but most are the size of a toddler's feet. Here and there, small rolls of clay dot the earth. Phallic symbols, experts inevitably claim, but they may just as well be the kind of absentminded leftovers every sculptor forms and then discards again when working with clay.

The bison were created in this utterly unlikely place some fourteen thousand years ago, deep in our prehistory. It was many millennia before the first writing script would be invented, let alone climbing gear or electric lighting. We can only guess at their purpose. There are no witnesses, no statements on record. Many experts see the bison as religious symbols, as they invariably do when it comes to cave paintings and other primordial works of art. To them, the animals form the centerpiece of an altar to fertility, a center for religious rites. But on closer inspection, there is nothing in the scene to suggest that.

For starters, why would people go to such trouble to hide their shrine? When shrines are hidden, it is usually because of religious persecution, but that was still a long way off. Or it may be because the rituals performed in them are secret, for the eyes of the initiated alone. So children tend to be unwelcome, if only because they cannot keep secrets. And yet, most of the footprints in the bison chamber of Tuc d'Audoubert are unmistakably child size.

What is more, a hidden sanctum would surely have been used on a regular basis over a period of many years. Rituals derive their strength and meaning from tradition, from staunch, meticulous repetition over a long period of time. The acts that make up a ritual must be performed "properly," in exactly the prescribed way and in the prescribed order, the way they always have been. If nobody can remember who came up with them or how, so much the better. The most powerful rituals are those that seem to have existed "forever," unchanged. So if the bison were part of a religious ritual, then a great many different people would have descended on the room on a great many occasions. The soft clay floor of the cave would have been covered in footprints of all shapes and sizes. But that is not the case. And regular users would have brought torches with them and lit fires, smothering the ceiling and the passageways leading to the chamber in a pitch-black layer of soot. But they are relatively unscathed.

That is not all. Although the two animals seem to represent a bull and a cow, the composition does not suggest anything like a mating scene, as we might expect for some kind of fertility temple. Had that really

The bison at Tuc d'Audoubert.

been the maker's intention, he was certainly skilled enough to have arranged things accordingly. Moreover, there used to be a third, much smaller animal near the forelegs of the rear bison. This was removed and transferred to a museum in Paris in 1912, when the bison were first discovered. A calf, perhaps? If so, the whole thing might just be a peaceful family scene, or maybe a birth.

All told, it looks rather like the artist operated of his own accord, bringing his children along for company or perhaps for lack of someone else to care for them. What is more, the bison chamber seems to have been simply abandoned afterward, lying forgotten for thousands of years. Luckily, we might add. For otherwise, the frail creations in it would not have been waiting, undamaged, to be discovered by the three brothers Bégouën, the sons of local gentry, one October afternoon in 1912.

* * *

So what on earth would have motivated someone, so many centuries ago, to worm his way into a pitch-black hole deep inside a mountain and mold some bison? And not just him, but also all those other creative souls who, between roughly forty thousand years and ten thousand years ago, adorned so many caves in France, Spain, and northern Italy with such a wealth of paintings and engravings?[1]

During this massive stretch of time, there must have been thousands of painters and engravers. In France alone more than 130 decorated caves and hideaways have been discovered. Judging from their skeletons and DNA, the artists were modern humans, anatomically indistinguishable from you or me. These hunter-gatherers are today generally known as Cro-Magnon people. The name derives from Abri de Cro-Magnon, an overhanging rock shelter near the village of Les Eyzies in southwestern France. It was here that, in 1868, the first fossilized human skeletons were found, in something resembling a modern grave.

"Cro-Magnon" is not a particularly meaningful name. They are not a separate subspecies, nor did they have a specific, well-defined culture. The name was coined mainly to distinguish them from the Neanderthals who shared their hunting grounds in Europe for thousands of years. The Neanderthals, who also owe their name to the place where their fossilized remains were first unearthed in 1858, in the German state of North Rhine-Westphalia, probably were a separate species. Their taxonomic status has long been a bone of contention, but recent genetic research indicates that at some point between 500,000 and 350,000 years ago, the Neanderthals split off from the rest of the hominid family and went their own way along the evolutionary path.

Neanderthals were around the same height as modern man, but they had a much sturdier build, with a heavy brow ridge, a low forehead, and a receding chin. But they were not exactly thick brutes. In fact, their brains were some 15 percent larger than ours are today. They made and used stone tools, and seem to have worn decorative paint and strings of beads. They may even have buried their dead in a similar manner to modern humans. How they got along with the Cro-Magnons is unclear, although DNA research indicates that there must have been at least some interbreeding. Only that could explain the fact that a small number of Neanderthal genes survives in each of us. But whether the two species mingled peacefully or resorted to raping and pillaging, we will never know.

An equally murky question is how the Neanderthals vanished from the face of the earth some twenty-eight thousand years ago. Were they exterminated by the Cro-Magnons? Did they lose out in the competition for the available resources as the last ice age neared its peak? Or is there some other explanation for their mysterious disappearance?

One possible account can now be discarded with confidence. The low percentage of DNA we share with the Neanderthals makes it highly unlikely that they simply blended into the mass of Cro-Magnons through regular interbreeding. The only other thing of which we are fairly sure is that they were not involved in creating any of the cave art discovered to date. A Neanderthal is thought to be responsible for a single crosshatching pattern found in Gorham's Cave in Gibraltar in 2014, a kind of self-expression not inconceivable in a species that adorned their bodies with ochre and beads. But to speak of "abstract art," as the discoverers did, would seem to be jumping the gun. The Neanderthals lagged far behind the Cro-Magnons in this respect.

The Cro-Magnons are thought to be the earliest representatives of our own species, *Homo sapiens sapiens*. They first roamed Europe around forty thousand years ago, or at least those parts of it not rendered uninhabitable by the ice age. When the ice thawed, they seem to have followed the reindeer they hunted northward, where they disappeared without a trace. By ten thousand years ago, they were gone—and so too was their practice of cave painting.

Meanwhile, they had built up an impressive oeuvre. Many of the paintings are of great beauty, testaments to the skill of their creators. By far the most prominent among these early cave "galleries" are those in Lascaux in southwestern France and Altamira on the northern coast of Spain.

The cave at Lascaux, a sleepy village on the left bank of the Vézère River, not ten miles from where the first Cro-Magnons were found, was discovered in 1940 by four teenagers. A few years before, a great old conifer had fallen over near the village, revealing a mysterious hole in the ground where its roots had been torn out. Curious, the four youngsters decided to go exploring, armed with little more than a flashlight. Clambering and sliding down the narrow hole, they stumbled into a great subterranean space whose walls, to their amazement, were covered with painted bulls.

The original entrance had collapsed thousands of years before, sealing off the age-old paintings from the outside world. A good thing, too, for otherwise little would have remained of the fantastic bulls, bison, deer, horses, and a unique hunting scene. In the first few decades after its discovery, the cave saw only a modest number of visitors. People were preoccupied first by World War II, then by the reconstruction of Europe. But by the early 1960s, they were showing up in droves, the moisture and acids in their breath and sweat visibly eating away at the paintings. To prevent further deterioration, the cave was closed to the public in 1963, and Lascaux II, a replica of its most important parts, was built half a mile up the road.

Lascaux was far from the first cave to be found with paintings dating back many thousands of years. But a mere sixty years prior, this was unheard of. All that was known to exist were a few engravings on the walls of French caves, presumed to date from the last ice age. This changed in 1879, thanks to an amateur archaeologist by the name of Don Marcelino Sanz de Sautuola. On a whim, he decided to take another look at an old cave on his land, some twenty miles east of Santander, in the Basque region of Spain. He had blocked off the cave years before, having found nothing of interest in it besides some stone tools. This time, he took his young daughter Maria with him. She followed him into a passageway so low that, although she could walk, he had to go in on all fours. "*Mira, papa: buyes!*", the little girl called out, pointing toward the ceiling. "Look, father: oxen!" Rolling onto his back, the Don saw that she was right. There they were, right above him: two dozen richly colored bison, several boars, a wolf, a pair of horses, and a handful of deer.

Don Marcelino knew right away: nothing even remotely like it had ever been found before. It was not just the rare beauty and elegance of the paintings. Nor was it the clever way in which the creators had

Don Marcelino Sanz de Sautuola.

used the natural shape of the rock to create a three-dimensional effect. What made this find unique was the thin film of salt covering the paintings, a telltale sign of their extreme age. Brimming with excitement, he proudly announced his discovery to the world archaeological community, only to be given the cold shoulder. "*Ancient*? Codswallop!" most experts proclaimed.

Truth be told, his opponents had reason to be skeptical. The find was so unprecedented that it was only natural to be wary of jumping to conclusions. But some, with far less respectable motives, began clamoring about outright fraud. Don Marcelino was subjected to a storm of innuendo, defamation, and ridicule, which only abated when, over the years that followed, one painted cave after another came to light across the border in France. Two decades would pass before Émile Cartailhac, the leading French authority on prehistory and one of the Don's fiercest opponents, finally acknowledged that he had been wrong about little Maria's find. His essay, entitled "Mea Culpa of a Sceptic," was an almost unprecedented scientific act of contrition, a testimony to Cartailhac's

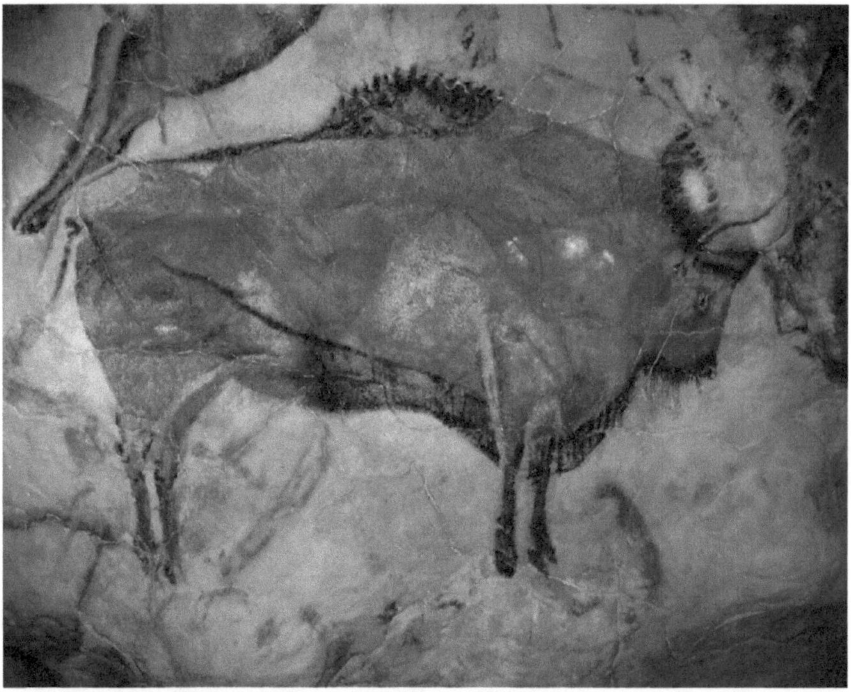

One of the oxen Maria Sanz de Sautuola saw on the ceiling at Altamira.

courage and integrity. But it was too late for Don Marcelino: he had died four years earlier.

Finally, the archaeological community accepted that, many thousands of years ago, people had walked the earth who were skilled and talented enough to produce paintings of astonishing artistic quality. But who were they? And what was the purpose of their work?

The line of thought that pervaded much scholarly research on the Cro-Magnons and their art went like this. Judging from the quality of the work, the painters had to have been well-trained, highly skilled craftsmen. So the Cro-Magnons must have been intelligent, culturally sophisticated, and eloquent modern humans. They must have passed on the knowledge and skills needed for engraving, drawing, and painting from one generation to the next. What is more, they must have been religious folk. They worshipped the animals they hunted as gods who had to be placated, and that was what their paintings were about. Or else they were a way of symbolically "catching" prey in ochre and soot. Alternatively, the images could have been part of some initiation rite that turned boys into hunters. The child-size footprints in the bison chamber in Tuc d'Audoubert seem to suggest as much. And invariably, the many signs and symbols surrounding the images were given some sort of sexual meaning. For instance, the many so-called "female signs," consisting of a V-shape or circle divided by a straight line, were readily assumed to represent vulvas. And single lines, just like the innocuous little rolls of clay at Tuc d'Audoubert, must surely be penises.

It comes as no surprise that the resulting picture of the typical Cro-Magnon closely resembled the self-image of Western man of the late nineteenth and early twentieth centuries. At that time, the visual arts as a profession was firmly rooted in the relationship between master and pupil, relying on formal training and lengthy apprenticeships, on sound knowledge of materials and finely honed skills. Painting and

Two views of the Cro-Magnon. On the left, a reconstruction from the 1922 bestseller *The Outline of Science*. The caption described him as a "representative of a strong artistic race." The prevailing view was of the perfect cross between a thinker, a respected artist, and an ascetic. On the right, the 1970s perspective: Cro-Magnons as primitive, sexually uninhibited humans. This 1971 film poster shows actress Julie Ege playing a Cro-Magnon in *Creatures the World Forgot* (photo Hammer Films).

sculpting were hallmarks of sophistication and culture, and the work of many contemporary artists had religious overtones. Organized religion also played a prominent and conspicuous part in daily life. Bishops, ministers, parsons, and a host of other ecclesiastical authorities had fingers in every pie conceivable. Moreover, statues and images played key roles in Catholicism, which was firmly entrenched in those parts of Europe where the cave paintings were found.

During the Victorian age and for a long time thereafter, sex and eroticism were at once both utterly taboo and a constant obsession. This was partly due to the Church's stifling grip on society. The unholy trinity of narrow-minded conjugal ethics, a lack of effective prophylactics, and the threat of incurable venereal diseases like syphilis and gonorrhea made matters worse. And then, as if to further fan the flames, Freudian psychology appeared on the scene. No sooner was a psychological phenomenon identified than it was explained in terms of lust, desire, and sexual frustration. In sophisticated circles, tall tales of the alleged sexual prowess and urges of the "savages" reached epic proportions. And so, quite naturally, the standard Cro-Magnon became a cross between what was then called a "native" and a post-Victorian Western European, clad in bear fur, a figure created in the image of those who had discovered, studied, and described him.

* * *

Was this image even close to what our artistic ancestors were really like? Not likely. Thankfully, the practice of interpreting any and all "female signs" as vulvas and single lines as penises has gone out of fashion. But there is much more that militates against the usual explanation of the cave paintings as having a religious, ritualistic nature.

We know that the staple prey of the Cro-Magnons were reindeer. If the paintings were meant to serve as some sort of hunting magic, a symbolic act of "catching" their prey on the cave walls, then we would expect mostly reindeer to be depicted. And to be sure, we do come across them occasionally. But much more frequent are images of animals that were too big or too dangerous for the Cro-Magnons to hunt—powerhouses like bison, tigers, mammoths, rhinoceros, all of them bruisers that were more to the tastes of the heavyset, virtually indestructible Neanderthals. But the Neanderthals did not paint.

Second, there are next to no images of wounded or dead animals, which you might expect of hunting art. Indeed, signs of any kind of suffering or violence are extremely rare. The one exception is the famous hunting scene at Lascaux, showing what seems to be a dead hunter, a spear, and a mortally wounded bison, its belly ripped open and its entrails spilling out. There is only one reasonable explanation for this lack of hunting imagery: the cave paintings had nothing to do with hunting rituals.

Equally unwarranted is the idea that the Cro-Magnons hoped, by depicting fearsome predators, to somehow appropriate their strength and cunning. This is a common superstition even today. Many people in Southeast Asia, for instance, believe that ingesting ground rhino horns and dried tiger penises enhances a man's virility. But if the Cro-Magnons hoped to enhance their hunting skills by transferring the animal's hunting power to its symbolic "captors," they must have had very strange ideas about hunting indeed. The animals pictured on the cave walls are largely bison, aurochs, mammoths, rhinos, boar, deer, and horses, all herbivores. None of these hunt.

So were the Cro-Magnons symbolically trying to ward off dangerous animals? That would explain the bison and the aurochs, the bears and the tigers, all of them creatures one did well to avoid. But it does not account for the multitudes of reindeer, horses, and other game that posed no real threat to the Cro-Magnons.

Spells and charms are a dead end. Sorcery was no more on the painters' minds than was religion. What all these pseudo-explanations for the artistic fervor of the Cro-Magnons have in common is that they rely on the existence of long traditions—traditions that would have to have been maintained for so many generations that the longest continuously existing civilization on earth, imperial China, lasted but a blink of the eye by comparison. But it is precisely the history of imperial China that reveals why it is highly unlikely that such long traditions ever existed in Cro-Magnon times.

* * *

Traveling through China for the first time can be a bewildering experience for a Westerner. It is a barrage of surprises, misunderstandings, and incongruities.

On the one hand, China is the longest continuously existing nation on earth. The empire was founded in the second century BCE and held out right up until 1912. It was a treasure trove of wisdom and sophisticated civilization, a land of the arts, high culture, and regal splendor. It was

China that laid the groundwork for the principles of decent government and legality. It was China that established the world's first professional civil service, while on the other side of the globe marauding Vikings were laying waste to large parts of poor, disorganized Europe. And it was the Chinese who invented paper, gunpowder, and the compass. China was a world unto itself, and the land is still steeped today in a sense of its own great past.

On the other hand, it is pervaded with the Maoism that held the country in its grip from halfway through World War II until the 1980s. Not too long ago, China was the emaciated nation of the iron rice bowl, torn to shreds by catastrophic civil wars, only to be traumatized still further by the atrocities of the Cultural Revolution. It remains a highly militarized, repressive, and closed society, a collectivistic anthill with little room for individual identity and self-expression. The eruption of raw capitalism in recent years has brought huge wealth to a happy few and moderate prosperity to some urban parts of the population, but it has also spawned outrageous excesses, bizarre abuses, and horrific pollution.

In all this time, however, one thing has never changed: the Chinese have great respect for the wisdom of old age. The greatest Chinese philosophers, Lao-tze and Confucius, were contemporaries of the earliest Greek thinkers. Although they were long dead by the time Socrates would usher in the golden age of classical Greek philosophy, they are still much venerated today. And at the age when European and American politicians generally retire, contemporary Chinese officials are usually just hitting their stride.

Westerners, too, have great respect for what is old, but in ways that have nothing to do with wisdom or experience. What we find admirable in great age has to do with products of culture, not people. We prize old buildings, monuments, works of art, books, dances, festivities, religious ceremonies, and so on. We see them as a sort of testimony to the past, tangible memories of days gone by. And we are less realistic than we are nostalgic, clinging to these objects to wallow in how things were better and more beautiful way back when. The rarer these relics are, the more we value and protect them. Moreover, this is a hobby we acquired fairly recently. It is part of a system of values that began to develop when the achievements of the ancient Greeks and Romans were rediscovered during the Renaissance, and reached its peak in the nineteenth century.

All this means nothing to the Chinese. They will destroy and repurpose historical artifacts without batting an eyelid. They routinely repair centuries-old temples using reinforced concrete, a mortal sin by Western standards. They bulldozed large swathes of Beijing's picturesque inner city like so much junk to accommodate the 2008 Olympics, and many more architectural and cultural heirlooms meet a similar fate all over China every day. And whether it is Louis Vuitton handbags, DVDs, or works of art, most Chinese could not care less about our much-treasured distinction between "the real thing" and fake.

Lest you might think that this kind of behavior is typical of the exotic Chinese, their way of dealing with the past has actually been the norm the world over as far back as we can think. When invading barbarians laid waste to the cities of the crumbling Roman Empire, these cities were rarely rebuilt. Instead they were left to rot, and ransacked for building materials to make new towns a few miles down the road. To this day, after wars, revolutions, and regime changes, we lose no time destroying the symbols of the old order. After Nikita Khrushchev denounced Joseph Stalin in 1956, countless statues and portraits were hurriedly destroyed or spirited away. His deeds were struck from school textbooks, streets named after him and his cronies renamed. Later, when the Soviet Union finally imploded and the Iron Curtain came down in 1989, the statues and portraits of Vladimir Lenin went the way of Stalin's before him. Recent years have seen efforts to reclaim both figures from the rubbish dump of unsavory historical figures in Russia, the east of Ukraine, and other places that were once part of the Soviet Union.

In Baghdad in 2003, the invading American forces were all too eager to dispose of the imagery of the fallen leader, Saddam Hussein. Rather than wait for the locals to get around to it, they staged a "spontaneous" iconoclastic riot, surreptitiously tearing down Saddam's greatest statue themselves. More recently still, the barbaric fanatics of the Islamic State, following the example of the Taliban in Afghanistan, have been butchering everything they can get their hands on in Iraq and Syria: in addition to thousands of lives, they have viciously destroyed countless ancient statues, artworks, and precious manuscripts.

Such fanaticism is typical of religiously inspired revolutions, especially if monotheism is the name of the game. Earlier followers of Mohammed razed churches and temples, then built their mosques on the rubble. The Bible is brimming with stories of the destruction of so-called idols, and early Christians, too, used to build their churches on the ruins of older heathen places of worship. As the burden of

Christianization passed from the hands of missionaries like Willibrord and Boniface to those of kings and nations, the methods used to eradicate any sign of older customs and cults became ever more ruthless. As late as the sixteenth century, Protestant rabbles went on devastating rampages in Catholic churches around Europe.

With this in mind, let us return to the Cro-Magnons' cave paintings. There were not all that many Cro-Magnons around, and they lived through an era twice as long as modern civilization, counting all the way from the invention of agriculture and the very first permanent settlements to the present day. So it stands to reason that many of the decorated caves must have been used, abandoned, and used again many years and many generations later by people who had no memory of or ties whatsoever to the previous users. If any given set of newcomers considered the paintings to have some religious or mystical purpose, we would expect them to have destroyed these scenes and symbols to make way for their own. But nothing like that seems to have happened. Not a single cave painting shows any sign of having been damaged or destroyed on purpose. Some drawings and engravings are superimposed on one another, but this does not seem to have been done with the aim of censoring the older images. So this, too, militates against a religious interpretation of Cro-Magnon cave art.

* * *

The most compelling reasons for doubting that the cave paintings have anything to do with tradition, be it religious, artistic, or any other kind, are numerical. From the vantage point of our overpopulated, frenetically busy world, it is all but impossible to imagine the sheer insignificance of our species for most of its history. Some seventy-three thousand years ago, the fate of humankind hung by an extremely thin thread. Mount Toba, on the Indonesian island of Sumatra, had just erupted. It was the most devastating eruption bar one in thirty million years, spewing out enough magma to fill a quarter of all the lakes on earth today, plus an unimaginable twenty-eight thousand cubic kilometers of ash. During the volcanic winter that ensued, the world population dropped to somewhere between two thousand and ten thousand. *Homo sapiens* was on the very edge of extinction, and our bodies still testify to the consequences. Since we are all descended from that tiny pool of survivors, the amount of genetic variation among modern mankind is unusually low. It is far lower than among even our closest relatives. The chimps and gorillas in their tropical rain forests seem to have been blithely oblivious to the volcanic winter that proved so disastrous to our forebears.

By the time the cave painting era began in earnest, some thirty five thousand years ago, the human population had bounced back to around one million. By the end of the era, around ten thousand years ago, this figure had grown to roughly ten million. For simplicity's sake, let us suppose that the population growth was perfectly regular, and the average number of people alive during the twenty-five thousand years in between was five million. A few exceptions aside, cave painting remained a specialty of Western Europe, despite the fact that people had already colonized the vast majority of the inhabitable areas of Africa and Eurasia, from the tundras of the north right down to the Cape of Good Hope. Given all this land to spread out on, it is not unreasonable to assume that Western Europe was inhabited by one-fifth of the world population. Perhaps half of these people lived in areas where cave painting occurred, amounting to an average population of roughly half a million.

Now suppose that on average Cro-Magnons produced their second child around the age of 20, which would suffice to keep the size of the population constant. This would mean that, spanning the entire Cro-Magnon era, about 1,250 generations of half a million individuals lived in the cave-painting areas, adding up to a grand total of 625 million people.

If the cave paintings really were the product of a continuous artistic tradition, in which knowledge of symbols, materials, and techniques was passed on from one generation to the next, then hundreds of engravers and painters must have been trained in each generation. There would have to have been at least enough to ensure that the death of an individual master would not put paid to the tradition in any given region. Say we allow for five hundred trainees per generation, the equivalent of one in a thousand people. Considering the Cro-Magnons' way of life, this would make artists very few and far between. As hunter-gatherers, they lived in groups of no more than a hundred people, one hundred fifty at a stretch. So most Cro-Magnon societies would never have even heard of a cave-painting artist, let alone have one in their midst. On the other hand, five hundred artists per generation over a period of twenty-five thousand years would surely have produced a staggering amount of work. Even if the average artist created only one serious piece of wall art, we are talking about more than 625,000 paintings and engravings.

Less than one percent of that number has been found. That is not much, given the number of caves that have been pored over with a fine-tooth comb. Even taking into account the fact that many works will have weathered away due to erosion, moisture, and fungi, and some caves may have collapsed altogether, that is still an awful lot of missing art.

And there is more. Whenever people are trained in something, they experiment. They try their hand at different techniques; they practice and play around to hone their burgeoning skills. So in the historical record we ought to see a good number of trial versions, rough sketches, and studies for every serious product. But these are nowhere to be found. Of course, some of the works we know of today might actually be trial versions fashioned by apprentices. But if so, this only increases the relative paucity of the final works found in proportion to the numbers that must have been produced. In any case, the number of paintings that might just qualify as trials and exercises is far too low to be taken seriously. So where did these artists-in-the-making practice? Could they have trained only in the sand and mud, so that their dabbling was quickly lost to the wind and rain? For people being trained to work on stone, this seems highly unlikely.

What is more, how could a tradition that managed to survive in the brutal environment of the ice ages, unbroken for over twenty-five thousand years—ten times as long as the Chinese and the Judeo-Greco-Christian civilizations—suddenly vanish without a trace? What could have changed all of a sudden that made the Cro-Magnons stop doing what they had been doing for so long? Are we really supposed to believe that, after more than a thousand generations of stubbornly sticking to their guns under probably quite trying circumstances, they just put down their paintbrushes and dissolved into the mists of time?

All of this becomes much easier to swallow if we accept that this was not a tradition in the usual sense of the word. The cave art came into being because now and again someone happened to paint a picture on his own initiative. Perhaps he stumbled across a work that was already there and felt inspired. Perhaps he was a lonesome individual with an unruly mind, or perhaps a whole group decided to get on board. Usually the idea would have been to create a single, isolated image, while at other times a more complex scene was planned, like the great panorama of the Grotte de Niaux.

* * *

The secret to the Cro-Magnons' knowledge of pigments and painting lies in personal adornment. Quite unlike the occasional, infrequent act of cave painting, decorating the body with pigments and mud probably was a widespread and continuous tradition. So, too, was the manufacture of small ornaments, from decorated bone, stone, and ivory, and doubtless also wooden figurines to beads made from teeth, shells, and other materials, all used as finery. These sorts of artifacts have been found virtually everywhere, including in most of the painted caves. And it is reasonable to assume that people who had pigments like ochre and soot on hand, and were willing to spend a lot of energy decorating their own bodies, made active use of those pigments. For one thing, it was a lot easier and faster to daub your skin with ochre than to produce a single string of beads. That involved first collecting a whole bunch of teeth, shells, or other objects of the right size and appearance, followed by the cumbersome task of drilling holes in them before finally stringing them all together.

Painting your own body does not require much in the way of civilization or culture, or even intelligence. It need not have any symbolic significance or other sophisticated meaning. A thin coating of mud, for instance, is an effective means of warding off mosquitoes and other vermin, as today's special forces are still very well aware. Adorning the body is not even a particularly human practice. Père David's deer, for instance, hang great masses of grass in their antlers to impress females as well as the competition. Bowerbirds build colorful palaces with lots of bling to lure the ladies, rather like some men on the prowl advertise their virility by way of a Lamborghini and a flashy gold watch. Other bird species lace their feathers with sticks, ostensibly to make them look bigger and more intimidating.

In 2014, some chimpanzees in Zambia were reported to place a blade of grass in their ear, a gesture that can only be understood as a kind of fashion statement. The practice seemed to catch on, spreading to other chimps in the group. It is remarkable in two ways. Unusually, it does not have anything to do with the usual purposes of attracting a mate or scaring away competitors. And second, the wearer of the blade had no way of seeing what effect it had on his or her looks. It seems to be one of the very few true cases of a purely ornamental, nonutilitarian act among animals. An act of self-expression, performed simply for its own sake.

There is no doubt that the Cro-Magnons and other early modern humans, even Neanderthals, were capable of personal adornment. And such skills and customs are readily passed on from parent to child. So we have every reason to suspect that all these people were enthusiastic body painters who used red and yellow ochre, charcoal and soot, and whatever other pigments they could lay their hands on. Their bodies served as a scrapbook, and learning took place through simple imitation.

Unfortunately, the people investigating the floors of painted caves and hideouts were often sloppy. This makes it next to impossible to establish a positive link between the beads and other objects found in the topsoil and the paintings above them on the walls. But their mere presence at so many sites is a strong sign that the tradition of self-adornment existed among the cave painters as well. Traces of this tradition have been in the form of small artifacts dating all through the Stone Age, indicating that it must have been upheld generation after generation, from the very beginning of the Cro-Magnon era. It was this tradition that led to the development of considerable artistic prowess among the Cro-Magnons—some of whom then felt the urge to turn to a rock face and create a permanent testimony to their existence.

* * *

Before we can try to uncover what moved the Cro-Magnons, there are two other matters we need to clear up. First, what kind of mental toolbox does someone need in order to begin drawing and painting? And crucially: is language part of that toolbox?

It is temping to jump to conclusions here. After all, drawing, painting, sculpting, engraving, and whatever else falls under the umbrella of visual art are just as exclusively human as human language. Beyond the odd bit of body paint and beading, even the Neanderthals do not seem to have had much of a taste for visual expression. So we tend to see the Cro-Magnons as very much "like us." But just because art and language are both part of the mental package that is typically human does not mean that one necessarily implies the other.

In fact, on closer inspection there seems to be no direct connection at all between language and visual art. The things required to make art have nothing in common with the things needed for human language. For starters, flexible hands and wrists are a must, along with highly developed, finely tuned motor skills. So animals, with the possible exception of the great apes, stand little chance in the visual arts department for purely physical reasons.

And that is before we even get to the mental requirements. Imagination is needed, for one thing. The mainstay of our imagination is our ability to recognize the semblance of a creature or an object in what is no more than an abstract configuration of lines and patches of color. Think of our stubborn inclination to see objects or faces in the clouds, the moon, and even cracks in the ceiling. Animals have no talent at all for this, nor any interest whatsoever. This is so much the better, perhaps, for an ability like this can be a mixed blessing in evolutionary terms. If a mouse were to see imaginary cats everywhere, it would die of fright and exhaustion before long. More advanced mental powers may not always be conducive to survival.

The young of certain bird species do seem, at first glance, to have something resembling a visual imagination. In a famous series of experiments, the renowned pioneer of ethology Niko Tinbergen fashioned a piece of cardboard like the silhouette of a bird of prey, with a short neck and broad wings. When this cardboard cutout hovered over the chicks' nest, panic ensued. However, it would be hard to call this imagination as we think of it. The chicks reacted *only* to the silhouettes of birds of prey, not to any other shapes. Clearly, over time evolution has selected individuals that respond most quickly and aptly to something that looks like a predator, for obvious reasons. Imagination does not enter into it.

By and large, animals share an astounding lack of interest in pictures, including images of other animals. Even dogs, famed for their territorial instincts, generally treat depictions of other dogs with supreme indifference. If a dog reacts to an image at all, it is purely to establish, after the initial surprise, that nothing untoward is going on. That done, it will lose all interest. Even a life-size photograph of its master will not cause a dog to jump around in excitement or wag its tail with delight. And who has ever seen a dog lick a picture?

Mirror images are a slightly different matter. Sometimes a dog will mistake its own reflection for an intruder and start barking furiously. But not for long. As soon as it becomes clear that the beast in the mirror is nothing to worry about, the dog will start ignoring it just as it ignores any other image. We tend to think this is very clever, that our precious Fido has figured out that it is only his own reflection. But Fido does not understand, not really. He has only decided that the image is harmless, worthy of little more than supercilious contempt.

This behavior toward images may seem odd, since animals perceive objects in their environment in the same way we do, by constructing a mental image on the basis of the impulses that reach their brains from their senses. What our eyes actually pick up may be light reflected off a real camel, a drawing of a camel, or even just some suggestive cracks in old plasterwork. Either way, all that hits the retina are rays of light. The camel itself is constructed inside the brain. In this respect, animals are just like us. The difference is that, unlike us, animals do not seem to construe anything meaningful from these retinal impressions.

It may be that, to spark their interest, animals need to receive information simultaneously from different sources. All a picture, photograph or sculpture has to offer is a visual likeness, and this alone does not seem to be enough. At the very least, it needs to be combined with movement, as with mirror images. This makes sense if you think about what some species do in the face of danger: they freeze. Stay perfectly still, and you have a good chance of escaping the notice of a passing predator.

This tactic can even work after the prey has been spotted, or even caught. Not long ago, I saw a frog escape what seemed like certain death at the jaws of a hungry cat. As it hopped across a sunny summer lawn, perilously far from cover, it was snapped up by a cat prowling around in search of a midafternoon snack. From my balcony, I watched it make a few desperate but futile attempts to get away, taking playful hits from the cat's claws all the while. Suddenly, it froze, lying motionless on the grass, no more than a whisker's breadth from the bared teeth of its captor. The cat lost its concentration almost immediately. After a minute or two of undecided hovering, the confused backyard bully decided to call it a day and slunk off in search of other adventures, leaving the still motionless frog untouched—and alive.

People are not put off so easily. If something in our field of vision freezes before we set eyes on it, we may not spot it. But if we do notice it, its stationary image is all we need to keep tabs on it. And whereas animals seem to need several types of information at once—shape and movement, for instance, or shape and smell—we can even make do with a less than complete visual image. In fact, we find incomplete, unusual, and sketchy images especially interesting. Children and adults alike can while away hours on all sorts of picture puzzles, visual illusions, and pictures in which other pictures are hidden. And we do not lose interest in pictures once we have seen them. Quite the contrary: the static visual arts are among our most highly valued kinds of artistic expression. Odds are that anyone standing in line at the Louvre to see the *Mona Lisa* has seen Leonardo's lady in one form or another many times before. Although the emotions fired by surprise and novelty quickly wear off, we derive pleasure from a painting or photograph on our wall, or a stone frog next to the garden pond, for years on end. Our pets miss out completely on that kind of happiness. They do not see the stone frog in the garden. They just see a stone.

Now we are getting to the crux of the matter. The point here is that, unlike animals, humans do *not* regard representations of people, animals, and whatever other objects occur in nature as the real thing. To an animal, anything resembling another animal is either *real* or it does not exist. But we immediately construct a mental representation, knowing full well it is not the real thing but a construct of our own mind. This is why an enormous photograph of a fierce-looking tiger on a billboard conveys to us everything we associate with tigers, but does not actually frighten us.

But what if that tiger were to move suddenly, or growl? We would get the fright of our lives. After all, our visual perception, like any of our other biological properties, is not entirely divorced from that of the rest of the animal kingdom. We still perceive things like other species do and respond in the same way they do. But evolution gave us something extra: a second mode of mental perception that truly sets us apart. We construct images in our minds of everything we come across, subconsciously remaining very much aware that these images are just that: images, not to be confused with the real thing in the physical world. So we do not ignore a painted seascape the way other animals do. We may actually be quite moved by it. But we know that we cannot drown in it.

Our fascination with film derives from a half conscious, half instinctive play on these two perceptual strategies. On the one hand, we know full well that a movie is just a movie. On the other hand, these "motion pictures" can be so realistic that they tickle the perception processes we share with animals, conjuring up all the lust, stress, and fear, all the flight and fight impulses that go along with them. The overwhelming size of the screen and the booming soundtrack also push all our animalistic alarm buttons. We hide behind our hands when things get really hairy, hearts in our throats—but we rarely flee for our lives altogether, thanks to that human perception process whispering at the edge of our consciousness that it is all just make-believe. Behold the secret behind Hollywood's success.

* * *

So we have the ability to tinker and tamper with the information we perceive. And this ability works in such a way that we do not end up with a false impression of the world around us. A good thing too; otherwise, we would act counterproductively and harm our chances of survival. The emergence of this ability, it is safe to assume, had a lot to do with the explosive growth of our predecessors' brains. Over a period spanning more than a million years, our brain capacity grew and grew. This, in turn, created a nice big home to accommodate our ever-richer inner world.

In the beginning, the burgeoning inner world of our earliest ancestors must have consisted mainly of memories of sensory impressions, just like the inner worlds of simple animals today (think back to our sea slug, for instance). An initial sensation caused neurons to fire, leaving patterns of excitation still smoldering after the initial sensation subsided. These patterns enabled the brain to react more quickly and strongly to similar subsequent impulses, and this in turn would have intensified the lingering afterglow still more. Thanks to the ongoing interaction between these two processes, these patterns became ever more deeply engrained in the brain, ever more persistent. Judging from the many different species that engage in learning behavior with long-term, sometimes even lifelong, effects, a jumble of neurons such as a brain seems to be well suited to keeping such patterns simmering for a very long time.

Apart from such sensory memories, in the course of evolution all but perhaps the simplest of species developed a stock of inborn concepts and abilities. How this happened, we do not exactly know. But we can safely assume there is no fundamental difference in the brain between these concepts and abilities on the one hand and acquired sensory memories on the other. What is important is that, more or less accidentally, some individuals were blessed with basic patterns of neuronal activation that were more conducive to survival and procreation than those of others. Thanks to selection, those patterns became part of the standard behavioral repertoire of the species as a whole. A pattern that caused the level of the stress hormone cortisol to rise whenever the image of a snake presented itself would naturally have been advantageous, but so too would one that enhances the coordination of muscle movements when walking.

So far this is all fairly standard, not unlike what happened in other species. But as our ancestors' brains became ever larger, they started filling up with ever more memories, concepts, and other stuff—far more than animals' brains could accommodate. What is more, the way our forebears' brains worked began to change as well. At some point, patterns of neuronal excitation were not only caused by sensory impulses from outside the body or signals from elsewhere in the body. Instead, they were also sparked by patterns already present and active inside the brain. Perhaps this was caused by electrochemical crosstalk between neighboring synapses, or by neurons, or by being part of more than one pattern. Thus, activating patterns awakened other, dormant patterns and perhaps even established wholly new ones occasionally. Eventually, all this gave rise to a self-sustaining chain reaction, a tangle of feedback loops of patterns activating, amplifying, or dampening other patterns quite independently of the outside world.

The emergence of these self-sustaining feedback loops must have been the start of associative thinking, and of consciousness, creativity, fantasy, and dreaming, because what underlies all these is the capacity to activate from within ourselves an endless variety of more or less related ideas, concepts, thoughts, and memories.

This development, like most others, was not entirely exclusive to hominids. Dogs dream too. All dog owners will have seen their sleeping quadruped friends twitching and moaning and pawing around in their basket. This is their inner world running wild, producing what seems like an orgy of joy and excitement, a world perhaps of unfettered existence, free of masters and leashes, but brimming with rabbits and turds and other doggy delights. Nevertheless, what their inner worlds are capable of is miniscule compared to ours. And the same goes for apes, dolphins, and elephants, the brainiest intellectuals of the animal world.

Our new, self-activating inner world must have come into existence in fits and starts, little by little. To observers, it would not have made much of a difference for a long time. Among the few tangible early consequences were the tools our forebears began to develop around two million years ago. But that is about it. Over hundreds, even thousands of generations, little changed in the way hominids lived their lives and saw their world. Nothing points to any unusual activity in the brain or any active realm of thought.

So intellectually, their lives were almost completely empty. But physically, by the time the Cro-Magnons arrived on the scene, the hominids had developed into *Homo sapiens*, anatomically indistinguishable from us and mentally a lot more sophisticated than old *erectus*. In fact, the species had arrived at a crucial junction.

The inner world of the Cro-Magnons must already have been independent enough of direct sensory stimulation to have acquired that typically human trait of being able to "see" something in almost any pattern of dots and lines. Their brains must have been bubbling over with spontaneous activity, their minds bursting with every kind of thought and image imaginable. Chances are that this unruly, unstructured torrent of thought is what drove some Cro-Magnons to making images, perhaps in an attempt to literally pin down and give order to some of what went on inside them. Many of their cave paintings make use of the naturally occurring textures, fissures, and bulges of the rock face, as if these conjured up thoughts and mental imagery of the animal to be portrayed, more or less the same way we can discern camels in cracks in the ceiling. Painting what they saw in their mind's eye may in turn have stimulated the further development of their inner world, just as artistic expression contributes to the development of our own brains. Like us, our ancestors must have derived pleasure from the artsy outlet prompted by their inner world; otherwise, they would not have kept at it for so long.

* * *

Since there seems to be no direct connection between the visual arts and language, it would be hard to infer anything from the artistic qualities of the Cro-Magnons about their linguistic capabilities. Yet language and painting do not seem to be entirely unrelated. It is hard to imagine either faculty developing in the absence of the kind of inner world just discussed, which operates independently and spontaneously produces its own thoughts, symbols, and images. Perhaps less conspicuous, but just as important, is the role controlled throwing plays in drawing and painting.

Controlled throwing implies the ability to take aim, a skill children start practicing from a very early age, not by actually throwing things, but by pointing at them—yet another trait that is characteristic of human beings. We constantly point at things. Other species do so only rarely, if at all. Infants who cannot even sit up yet will tentatively extend an arm in the general direction of interesting objects, hand open and index finger outstretched, in a gesture halfway between clawing and pointing. By the time they reach their first birthday, their motor skills have improved such that they can point at things effectively, if still somewhat clumsily. Around the same time, they also throw themselves into yet another great human hobby: naming things. From those first few words, their vocabulary will ultimately grow to comprise tens of thousands of items. Like Adam and Eve before them, all toddlers go around their own Garden of Eden, pointing and naming to their heart's content. "Train, train, train, papa!", they will exclaim from behind a frenetically outstretched little paw. Their world is positively filled with intriguing and enticing things, all of which need identifying and categorizing.

This naming phase is a crucial step toward a full-fledged language faculty. But it is not the only thing going on. At the same time, children are also subconsciously acquiring the specific grammars of the language (or languages) spoken around them. This enables them to flesh out the general grammatical scheme with which they were born, into a set of rules for building words and sentences that fit their mother tongue—and for understanding them. By labeling objects, events, concepts, and feelings, toddlers build the vocabulary necessary to give their sentences referential content.

The frescoes and painted ceiling of the Sistine Chapel in Rome are among the most famous works of art in history. One part of Michelangelo's masterpiece stands out in particular. It is the intriguing scene, right in the center of the ceiling, in which God gives life to Adam. Judging by the endless copies and the number of satirists, cartoonists, and even filmmakers it has inspired, people seem to be strangely attracted to it. Remember Stephen Spielberg's 1982 blockbuster *E.T. the Extra-Terrestrial*? Part of the fresco's appeal may have to do with the way Michelangelo has Adam receive his precious gift. The Creator does nothing more than point at the tip of Adam's left index finger, all but touching it. So spiritually loaded, so meaningful and impressive is mere pointing. But why is pointing so important? We can only speculate, but it may be because pointing unites two essential human traits. It does not seem unreasonable that our irrepressible urge to name things may have followed from our equally irrepressible habit of pointing at things, which in turn arose from our controlled throwing ability. If so, then throwing was of fundamental importance for the development of our language faculty in two separate ways. It provided the mental machinery needed for the high-speed crunching involved in coding

Adam receives the gift of life on the ceiling of the Sistine Chapel.

and decoding sentences. But it also lies at the bottom of our penchant for acquiring ever more words.

From here, it is only a small step to the human tendency to create images. Although most of us tend to lose interest as we age, there is hardly a child to be found between around three and ten years old who, given half a chance, will not produce one drawing after another. These children are smack in the middle of the process of language development. They will draw anything from happy scenes full of rainbows, mommy and daddy, and Fido the dog to images of pure trauma, reflecting the horrors the less fortunate among them have been through. But although they generally draw people and things from their surroundings, what they never do is try to draw to life. No little girl will ask her brother to sit for a portrait session, or even so much as look at him before she sets to work. Nor will she go out and sketch a scene from life, like adults do. What children portray is their inner world. Perhaps this urge to draw is yet another expression of the human need to identify and name, to create order and gain control of all the new impressions pouring incessantly into a child's inner world.

* * *

The Cro-Magnons probably had all the biological building bricks to construct a modern mind, including that all-important language faculty. But for all their drawing and painting, their repertoire remained quite small. Most of their works are of animals, but not just everyday creatures. The Cro-Magnons had a clear preference for spectacular beasts, the more awe inspiring and dangerous the better. Interestingly, they seem to have painted reindeer, one of their staple foods, more often during periods in which these animals figured less prominently in the Cro-Magnon diet. They rarely painted small animals, plants, and fish, the bread and butter of the Cro-Magnons, so to speak. And they almost never painted people or their possessions.

All this points to a mind that is capable of wondering, but has little grip yet on what goes on inside it. The Cro-Magnons seemed to have painted whatever amazed them. Barring a few exceptions, that meant the weird and wonderful things they observed in the world around them—things they really *saw*. Apparently the uneventful, the usual, and the mundane did not make enough of an impression. This would explain why reindeer were depicted more often when they were in short supply. Judging by their output, Cro-Magnons never consciously *saw* themselves or their peers, children included. This must mean that in the theater of their minds, they had not really become detached from the natural backdrop. They were still creatures with no sense of self, no self-awareness. They were not yet fully aware of their own individual existence, as separate, self-contained entities moving independently through their environment, in it but not part of it. And it is unlikely they could form sophisticated theories of mind, the capacity we have today to view the world from someone else's perspective. In this respect, they were not so different from chimpanzees.

But in other ways, and unbeknown to themselves, they were quite different from chimps. They had almost completely walked the long road toward modern humanity. Inside the Cro-Magnon mind, something was glowing on the horizon. There was no real individuality yet, no clearly defined ego. But two things suggest that there was a vague sense of Dasein, of *existing*.

On rare occasions, a human figure shows up in the cave paintings. The most famous example is the hunting scene at Lascaux, which is exceptional in more than one way. Violence and death almost never figure on the ancient cave walls in France, Spain, and Italy, but here we are confronted with a bison in the throes of a horrible death. It is still on its feet, stamping and snorting with rage. But its fate is sealed, for its entrails hang from its ripped belly like a gory tangle of ropes. Equally exceptional is the presence of a hunter. He is the victor, but will never have the pleasure of savoring his victory. He lies dead on the ground, fatally wounded by the horns of his adversary. His penis is fully erect, which some say is not altogether unusual in cases of sudden, violent death. But in contrast to the realism of the rest of the picture, the human hunter is a primitive stick figure, no more than a clumsy rectangle with stakes for arms and legs. He is there, but only in the most schematic way possible, more a scarecrow than a man.

This is one of only a handful of depictions of human beings, or even hybrids of man and beast, against thousands of images of animals. And yet the Cro-Magnon painters left their mark everywhere. They literally stamped their signature on the cave walls in the form of thousands of handprints, all over the Cro-Magnons' territory. There are positive prints, made by pressing a hand daubed in ochre onto a cave wall. But there are also many negatives, made by blowing a powdered pigment at a hand pressed against the wall. As a gesture these signatures are reminiscent, again, of pointing. It is as if in some vague way their creators were trying to identify themselves by marking their own presence, as though they were on the verge of becoming conscious of themselves.

There is something strange about these handprints. Some of them appear to be missing one or more fingers, either partly or altogether. With negative prints, a partly dismembered hand can be simulated by bending a finger or two double. But this would explain only some of the partial prints. There is no doubt that Cro-Magnon life was rough and that inevitably many a bone was broken. Yet it seems unlikely that the Cro-Magnons had so little regard for their hands that they routinely lost fingers entirely. One possible explanation is intentional amputation as part of some religious ritual. But besides being barbaric, this would surely have been utterly stupid. Why willingly mutilate the tools on which your existence depended? There are plenty of options that are less detrimental—why not just have someone pierce your nose, carve some scars on your torso, or even knock out a tooth or two, as still happens in some societies today? Quite apart from this, it seems unlikely that the Cro-Magnon mind had reached the level of sophistication required for such elaborate religious ideas and practices. We can hardly attribute an understanding of notions like "god" to people not even capable of discerning themselves as individuals, separate and distinct from the collective. How could they have conceived of a deity? Regardless of what the Bible says, it is man who creates the gods in his image, not the other way around.

Bonobos may hold the key here. They are known for being cheerful, horny hedonists, happy to resolve conflicts through frenzied lovemaking—a far cry from their cousins, the unforgiving, violent-tempered chimpanzees. But even these bon vivants of the animal kingdom indulge occasionally in nauseating torture practices. In the 1970s, Belgian biologist Jeroen Stevens figured out why some bonobos lack one or more fingers. At a zoo in the Dutch city of Apeldoorn, he witnessed a pack of five bonobos set upon a sixth and, to his horror, chew off its fingers. And they were not content with merely biting them off, like a dog or crocodile might. Instead, they literally gnawed away great chunks of the victim's hands, tearing off filaments of flesh with their teeth. If the good-natured bonobo is capable of such excesses, it is not unimaginable that the Cro-Magnons may have known similar punishments.

* * *

At this point, the Cro-Magnons may appear less familiar, less human than ever. By now, not much should be left of the cozy late-nineteenth-century image of the deeply religious noble savage with his close-knit family unit, infinitely more cultured and sophisticated than those coarse Neanderthals. The truth is, the Cro-Magnons were not like that. Nor were they like us. If time machines existed, visitors from the modern world would have a hard time communicating with them. Creatures that still lacked a full-fledged ego, to whom not just the terms but the very notions "I," "you," or "she" were utterly foreign, would not have had anything like a fully operational language faculty. In many respects, their society would have seemed closer to something like a chimpanzee colony or a bonobo tribe.

At the same time, the Cro-Magnons came very close to sprouting real self-awareness and consciousness. Their minds were teetering on the tipping point, right on the verge of splitting off from the two-dimensional scenery of their group and their environment to become truly independent, with all the dignity and loneliness that comes with it. If they went over the precipice, never again would they enjoy the unconditional security of being an integral part of a group quite so much. But the language faculty it came with would give them a far richer, infinitely more powerful inner world in return, with all the cultural gains it would generate.

Being right on the edge of true modernity, the Cro-Magnons must have already possessed a whiff of individuality, an incipient sense of self, and with that, some idea of the flow of time, of finiteness, of irreparable loss and death. After all, they did bury their dead, occasionally even with burial gifts. This is not something you do unless the deceased is in some immaterial sense still "there," still present someplace in your inner world. Equally, the Cro-Magnons were a hair's breadth from developing a modern, fully functional language faculty. As we have seen, they certainly had all the necessary building blocks. They were completely capable of controlled throwing, and they must have had the mental machinery that enabled them to sing and use rhythm. What is more, they already had a lively inner world. Their minds must have been raging constantly with half-baked ideas, feelings, and thoughts, including those that prompted them to start drawing and painting. But save for their cave art, they accomplished little to nothing over tens of thousands of years. So they cannot have had much control over what went on between their ears. More than anything else, structuring and controlling that gamut of thoughts and feelings would be the task of the language faculty.

It is not easy to imagine how a Cro-Magnon must have felt. No matter how hard we try to project ourselves into their minds, we cannot escape our modern self-awareness and individual sense of identity. We are very good at developing a theory of mind of the other, an image of what the world looks like from their point of view. But such a theory is always based on our personal experience of our *own* mind. We can never really know what it is like to be confronted by spontaneous snippets of thought all the time, without having the means to order and structure them in a way that makes sense. We will never know what it is like not even to realize that those sensations in and of the mind are *thoughts*, for to do so requires a certain level of self-awareness and the power of self-scrutiny. So the Cro-Magnon mind remains beyond our understanding forever—just as ours would have been to them.

In small children, we can still see something of what it must be like to exist without a true sense of individual identity, as an integral part of a collective and as one with the environment. Kids are great storytellers, and will regale their parents, grandparents, and anyone else willing to lend an ear with their adventures. Their stories differ from those of grown-ups in two ways. First, they make things up. They routinely mix facts with fantasy, but are utterly unaware of doing so. Second, they tell their stories as if their audience already knows everything about their lives. This can be completely bewildering. Where are we? When was this? Where did that animal come from all of a sudden? Who on earth is "he," and what about "the lady"? Children, on the other hand, cannot figure out for the life of them why you would need all that to be explained. They just do not understand that the people around them do not automatically share their own personal experiences. They think that whatever they see, hear, and feel is seen, heard, and felt by everyone around them. They consider themselves naturally transparent—part of a collective mind, if you will. Only later do children become aware that they are individuals, separate people in their own right, with minds that are walled off from everyone else's. Only then do they learn what it means to keep a secret, and catch on to what lying is all about. It is this Cro-Magnonesque mind-set, if you will, that explains why small children are notoriously difficult to interview and make extremely unreliable witnesses.

It is not unlikely that Cro-Magnons occasionally heard voices, just as some of us do today when our inner world is not entirely in check. These voices can be deceptively real—the brain can, of its own accord, produce the actual sensation of hearing an actual voice. And hearing them can be a frightening experience, whether or not the hearer knows they are not real. Not only because such voices often confront the hearer with the darker recesses of his or her own mind. The inability to silence voices that should not be there is equally unsettling, and undermines our sense of autonomy. Interestingly, young children hear voices more often than adults do, although this is not usually recognized as such. Instead we—and they—tend to say they have an imaginary friend. Only when a child is depressed, lonely, or otherwise disturbed do the voices sometimes take on the frightening, even malicious character more common in adults. This, too, may be due to their still-developing sense of identity as separate individuals. As long as children are to some extent at one with their environment, as long as there is not yet a stand-alone, isolated "I," voices and imaginary friends are not experienced as intruders. Likewise, Cro-Magnons would not have been much bothered by the voices they heard.

Lacking the control mechanisms that come with a full-fledged sense of identity and the thought management system we call language, the Cro-Magnons had no sophisticated way of separating real sensory information from memories and delusions. They must have lived in a state that we today think of as mildly psychotic. A psychotic episode is like a frightening, unstoppable gale raging through your inner world. To Cro-Magnons, this must have seemed normal, and the same would seem to hold for very young children. Again, the lack of a sense of personal identity may have made this less alarming than it sounds. Naturally, we can only imagine how it would have affected Cro-Magnons, but we are all familiar with the chaotic ramblings of young children. Far from being scared, they are perfectly happy during such episodes, except, of course, when there are monsters under the bed. Far from psychosis, parents tend to chalk these outbursts up to nothing more than a lively imagination.

Nevertheless, there must be limits to the strength of these inner-world storms. Neither children nor Cro-Magnons could survive, let alone grow and prosper, in the never-ending throes of mental turmoil. So what stops our kids going stark raving mad, living in a world positively brimming with novelty and things to learn? How can it be that they develop into stable, self-possessed adults? The answer seems to lie in the growth trajectory that takes us from infant to adolescent.

At the outset, a child's inner world, such as it is, is almost without controls. It is the language faculty, which enables us to form more sophisticated structures out of the raw contents of our inner world, that helps us to keep order in our minds. Starting from an innate core, this faculty takes the best part of ten years to ripen into its fully functional form. The speed of development varies from child to child, but all follow the same path, passing through the same stages, toward perfect command of their mother tongue—or tongues, as the case may be. And as the mechanism approaches its mature state, so too does the child's control over their inner world and their ability to manipulate its contents. In other words, the growing power of the inner world goes hand in hand with the development of the controls that keep it in check. This also helps explain why young children sometimes fail to register highly salient events, even traumatic ones. This is not a defense mechanism; they are simply not yet capable of processing and storing such experiences. And if certain aspects of an early trauma do register, there is a good chance the child will later end up on a therapist's couch in order to process them properly.

Cro-Magnons, however, did not get to enjoy the increasing order and control that accompany a developing language faculty. So why did not their unchecked minds drive them up the wall? One important factor may have been the relative emptiness of the Cro-Magnon world. Compared to later, more sophisticated societies, Cro-Magnons had very little information to process, store, and work with during their lives, and especially very little that was novel. For most of their lives, virtually nothing went on. There were no inventions, no new ideas. And given the lack of language, there were no stories to take in or tell, no dreams to share, no discussions or debates to be had, and very little learning. When it comes to creating sophisticated thought patterns, language is indispensable. So their inner worlds must have been quite inert. You could think of their minds as a big box of Meccano without the nuts and bolts. Combined with their absence of "self" and the fear of "losing oneself" that goes with it, most Cro-Magnons probably lived quite happily with their uncontrolled inner world, more or less like chimpanzees.

But not all Cro-Magnons did so, because unlike chimps, they had great, powerful brains full of machinery, and were on the verge of developing a new, completely different kind of mind. Every now and again, an individual must have come along who was just a little different, a tad more advanced than the rest. In them shone a faint light, making them more susceptible to the turmoil in their minds, yearning for some outlet to express the images careering through their inner world, urging them to create a different kind of feedback loop, one that included the physical environment. They satisfied this urge by picking out a surface and covering it with whatever they saw in their mind's eye, just as small children do (although since their minds were different from children's, so too was their subject matter). Why they so often chose such inaccessible locations to do it will forever remain a mystery. Perhaps the darkness and isolation made them feel safe and sheltered enough to give their mind free rein. Perhaps they felt closer to themselves there, better able to listen to their inner voice.

One thing is crystal clear. Whatever motivated the Cro-Magnons to paint, it was not religion. Nor was it a creative urge in the modern sense. It was not *art pour l'art,* art for art's sake, although it did flow from the desire to express the ruckus in their minds. It was their growing inner world beginning to bubble up to the surface. These were still simple folk, in whose brains a nascent language faculty was slowly coming together, bit by bit, a faculty that would enable them to create a whole new world. It must have been the painters who felt most keenly the first sparks of that process, urging them to seek out the quiet of a deep cave to express themselves to their heart's content. Shamans in the shadows.

Note

1. So far, only two reliable sites have been found outside Europe: the Apollo 11 Cave north of the Orange River in Namibia, which has animal paintings more than twenty thousand years old, and a complex of limestone caves near Maros, on the Indonesian island of Sulawesi, with ochre handprints and depictions of babirusa, or pig deer, dating back some forty thousand years.

6

Voices on the Plains

> A hawk was being harassed by a pack of crows. To gain altitude, they beat their wings like drowning swimmers thrashing about, then swivelled and dived as if to ram their opponent, squawking madly but always deliberately missing their target. At each feigned assault, the hawk adjusted the angle of its outstretched wings and glided on, surfing the currents of hot air rising from the rock and scrub beneath. It could easily have turned on its tormentors and gutted them with its great claws, but killing on the wing was alien to its species. For their part, the mob of crows might have attacked this competitor on their territory in earnest, flustering it enough to give one of them an opening to drive its spiky beak into the intruder's body, but neither was such behaviour programmed into their genetic code. It was thus a confrontation that neither protagonist could win decisively, and would go on and on until one or the other tired of the game and gave up.

This is the scene Italian police inspector Aurelio Zen witnesses in *End Games* as he relaxes in the shade of the Calabria mountains, his bread, salami, and wine spread on the grass before him. Zen, the hero of the books by the late British author Michael Dibdin, is just another policeman on a picnic, but his observations are astute: the behavioral repertoire of most animals is almost completely hardwired into their genes. It is inextricably bound to a rock-solid, immutable, innate program, not to their actual physical capabilities. Not only Zen's crows and hawk behave as Dibdin describes. All crows do, as do all hawks, always.

People are very different. There is no doubt that we, too, are endowed with a genetically determined behavioral repertoire. But for the most part, we live and thrive by constantly inventing novel ways of dealing with each other and with the rest of the world. However strongly Napoleon maintained that the secret to being a successful general was good luck, just as important is the ability to act unpredictably. Guile, subterfuge, and the element of surprise are the most powerful weapons available to man, that is, doing things *differently* every time we do them. This is because, when it comes to getting the better of other people, there is no fail-safe protocol, no standard procedure. A hunter will always be able to bag a deer by staying downwind and keeping mum—the same trick works every time. By contrast, a ruse against other people will only work as long as they do not recognize it.

This kind of creative thinking is beyond animals. They cannot imagine new things and will them into being. They cannot think outside the box or go about something in a completely novel way on a whim. They cannot change tack midcourse. And their behavior almost always stays well within the bounds of their genetically programmed repertoire; they have no way of exploring the limits of their capabilities.

Of course, animals sometimes do acquire behaviors that are not part of their hereditary endowment. Take the circus. Given enough training, elephants can learn to balance on a ball and tigers will even jump through flaming hoops. In dressage—an Olympic sport, no less—horses are taught to prance to music through a sandy pen in highly peculiar ways, also known as "horse ballet."

But for all their learning skills and their physical prowess, the point is that not one of these animals—or any other, for that matter—will ever come up with such ideas on its own. It is not in their natural repertoire, and that is that. With very few, very limited exceptions, animals will only show novel behaviors when coaxed or forced to do so by people.

In us, however, the urge to do things differently is almost irrepressible. We have a natural desire to put our capacities to use in all sorts of novel ways. Think of sports like tennis, soccer, golf, or cricket. All these involve highly convoluted systems of strict rules, and players bend over backward to obey them. Yet they do so for no good reason; there is no rational gain to this. The only reason for inventing and sticking to such abstract, largely meaningless systems is because we *can*—just for the fun of it, out of curiosity. The very fact that something might be feasible is reason enough for many people to have a go at it. And so people try to fly with self-made wings, or go bungee jumping, or spend weeks setting up a circuit of millions of dominoes only to watch them be knocked over in a matter of minutes.

The mere inkling that some mystery might be solved, some esoteric phenomenon explained, or objective achieved—the trick to making gold, say, or the secret behind flying—is incentive enough for some people to devote their lives to trying. We call these people scientists or inventors, explorers or buffs. Top athletes fall into this category as well. In essence, they devote their whole life to a specific, fairly useless physical trick. For the sake of being better or faster at something than everybody else, they are willing to sacrifice their general well-being, knowing all the while they will inevitably lose the fight sooner or later as age and the competition catch up with them.

Even the far less ambitious and competitive among us are never truly inactive. We all indulge the desire for new challenges by playing cards, doing crossword puzzles, or getting stuck in a Sudoku puzzle. Every new round is yet another small adventure. We explore unknown territories by perusing newspapers and magazines, reading novels, or surfing the Internet. Some fight blood-curdling battles in the virtual worlds of computer games. Whether it be ogling the television or building a seaworthy vessel from lollipop sticks, we cannot live without something to occupy our restless minds any more than a cat can forgo sharpening its claws.

In the absence of useful, pleasant, or even completely senseless tasks, boredom sets in. Days and even minutes drag on forever, and we get agitated, aggressive, and depressed. Boredom is largely a human emotion. Animals betray traces of it. Farmers are only too familiar with the things barnyard animals will do when cooped up and denied the chance to perform their normal behaviors. Horses start cribbing, pigs will chew off one another's tails, and chickens go in for bloody bouts of pecking. To us, this may look a lot like the wanton destructiveness of bored teenagers, but its causes are actually very different. It is a reaction to physical restraint; they are acting out of stress and frustration. Pigs and chickens are often packed in more tightly than their minds can handle, putting them in a state of constant panic. It must be like having alarm bells ringing inside your head all the time and being unable to turn them off.

Bored human young, in contrast, usually suffer from a lack of opportunity. They will complain of having "nothing to do" when what they really mean is none of the available choices are what they really *want* to do. Their inactivity is entirely a mental phenomenon; they are not physically cooped up in any way. But they suffer from it all the same, growing restive and aggressive.

This hardly ever happens in animals. Take elephants. In zoos, you can sometimes see them standing in their pen, slowly rocking to and fro, shifting their weight from left to right and back again. For want of more interesting input, they are moving to the tune of their own internal music, the beat of their hearts. They are probably none too happy, but there are no signs of frustration or restiveness. They are not bored in the human sense of the word.

Or think of dogs, which can spend hour upon empty hour waiting with what seems like superhuman patience for the return of their masters. As soon as they hear the key turn in the lock, they jump for joy and forget immediately about their past plight. They would not dream of avoiding or growling at their returned master, the way people kept waiting will mope and grumble when at long last the waiting is over. Nor will well-bred, healthy dogs express displeasure about having to wait by barking, howling, or peeing in their owner's bed. And they do not behave this way just for fear of punishment on the boss's return. Any misconduct they do display in the absence of the top dog, their master, stems more from fear of abandonment than the fact of being made to wait.

This is not surprising, since dogs spend most of their days in idleness anyway. A bit of poking around in their basket now and then, a stroll from the coat stand in the hallway to the kitchen door, some lengthy bouts of lying in everybody's way—that is pretty much the extent of their daily activity. If something exciting happens—say, the postman daring to approach the door—they are suddenly the very picture of bouncing, barking activity: Aha, fun! Thrills! "Woof!" But if excitement fails to materialize, that is okay as well.

In the wild, life is not so different. When animals are not busy finding food, they fill their time with a whole lot of nothing. With the exception of cows put to pasture for the first time after winter, useless exertion is almost exclusively the prerogative of the young. Otherwise, they know better than to expend energy for no good reason.

"Know better" is a bit misleading here. In reality, animals can never behave otherwise, for to get bored in the human sense of the word, you need to have a sense of time passing. To sense that something is dragging on, or even has any duration whatsoever, you have to be able to see time flow before you as if it were a river. And to do that, you must be able to step out of time and watch it from the banks, as it were. This requires the capacity to see yourself as distinct from the environment of you are part of. You have to have a sense of yourself as standing apart from the world, watching that world, including the river of time, go by. Animals do not have this.

Once you do have a notion of the flow of time, things start to *take* time. And they keep on doing so, because, having developed this notion, you can no longer turn it off. You cannot help but start measuring and apportioning time. You become aware of fleeting moments, brief spells, and long periods of time. You notice how long your urges and wishes remain unsatisfied and inconveniences persist. A notion of time makes us impatient, and it gives rise to boredom if it is not filled with external stimuli—or internal stimuli, of course: people who are thinking or daydreaming may seem to be doing nothing, but they certainly are not bored.

* * *

Although we respond to dullness in a very different way from other mammals, the effects of extreme stimulus deprivation on our health and psyche are similar. It is well known that lab rats kept in bare metal cages are less healthy, perform worse, and suffer from more stress than animals in so called "enriched" environments, with chaff on the floor and some structure to play with, climb on, or hide behind, and of course, some company—but not too much. Cage-dwelling pets like marmosets, guinea pigs, hamsters, and rabbits also do better in an environment in which they can more or less live as they would in the wild. All too often, however, their loving owners let them wither away in cramped, stimulus-poor cages. Today's much-pitied lab rat may in fact be better off; they are at least cared for by professionals and protected by regulations on housing and handling them. Zoos, too, have been the source of much suffering. Even today, some zoos still keep animals in isolated, cramped, and stimulus-poor conditions, making them incurably antisocial or just plain mad.

People are equally vulnerable to shortages of stimuli. Solitary confinement, preferably in a dark, bare room, is one of the torturer's favorite means of breaking the resolve of uncooperative prisoners. It does not take long for prisoners to lose their bearings, start hearing voices, and begin to hallucinate. Long-term isolation literally drives people insane. And it is not the loss of personal autonomy and decorum, nor even the extreme humiliation inflicted by their torturers that does the greatest harm to these prisoners' state of mind. Instead, it is specifically the lack of stimuli.

Something similar can be seen in nursing homes. People whose world has shrunk to a bed and a common room shared with a few mumbling geriatrics often begin to show signs of mental decline, especially if their eyesight and hearing are deteriorating as well. As external input decreases, their inner world and the world around them begin to blend into one, and they start to lose their grip on reality.

From the predicaments of prisoners and the elderly alike, we know that a minimum of input is indispensable if we are to keep our restless minds on the right track. This need may also explain what goes on in our minds when we dream. In our dreams, the laws of everyday reality are pushed aside. On the one hand, dreams are usually rooted in reality: they are filled with actual people and events from our waking experience. On the other hand, they are full of inconsistencies, a place where vague fears and worries fester, where warped memories and strange phantasms flourish.

Some of these fears and worries appear to be universal. Everybody seems to have dreamed at least once about being stark naked in public with no one noticing. We have all had the kind of dream in which a dozen people are coming to dinner, and we suddenly realize we have not done the shopping, but our normal shops have just closed so we have to take the bus to some faraway supermarket, but the bus will not come, and when it does it is actually a lift with a washbasin in it. This is the standard stuff dreams are made of, next to missed trains, unclimbable stairs, journeys leading nowhere, missing keys, goals that are always just out of reach, strange revolving doors and, of course, erotic adventures with partners who are absent, nonexistent, or out of bounds. Often we dream of things that occupy our minds in real life, steeped in a rich tapestry of bizarre plot twists and surreal mix-ups.

Dreams are phantasms in the most literal sense of the word. They may very well be caused by the fact that while we sleep, our brain is almost completely bereft of input from the outside world. It seems that even a relatively short period of deprivation of contact, combined with a relaxation of the reins of consciousness, is enough to let our inner world break out of its normal confines in a fairly spectacular way.

Dreaming is something we absolutely do not want to do while awake. Losing our grip on the guidance provided by sensations from the outside world comes close to what we call psychosis. This is when the external impulses that temper our thoughts and actions are drowned out by strong, uncontrollable images conjured up by our inner world. It has parallels with the self-induced loss of control that results from tripping and certain kinds of meditation.

To bystanders, tripping, the kind of mental adventure induced by psychedelic hallucinogens like LSD, psilocin, and psilocybin, may look like psychosis. If things go really wrong, it may also have similar effects. But tripping is actually very different. It is not a means of having your inner world override the input from the outside world, as happens in psychoses. Quite the contrary. It forces the dampening, interpreting functions of the inner world onto the back burner, giving free rein to external sensations. Pot has a similar, but far weaker effect. Stoned people experience colors more intensely, sounds more loudly, and sensations more strongly. In other words, everything is a bit larger than life. They can end up in fits of uncontrollable giggling or other mildly excessive behavior, but also unnecessarily anxious, fearful, paranoid, or depressed.

Some forms of meditation work the opposite way, by denying the mind its external input—it is not for no reason that some people find meditation akin to torture. Other ways of achieving this kind of ecstasy are practicing self-flagellation, abstaining from worldly pleasures, or even retreating to a hermit-like environment devoid of all stimuli. But even just concentrating on a single, self-produced external stimulus, like the "om" mantra or endless repetitions of the Lord's Prayer, is enough to send our inner world off on its own. Meditators do not experience the same effects as potheads, but may have all kinds of visions, up to and including chatting with the Holy Virgin. What these techniques and the use of psychedelic drugs have in common is that the hallucinations are conjured up by a person's willfully relinquishing normal mechanisms of mental control. So meditators can put an end to their ecstatic condition by breaking their concentration or seeking out a more stimuli-rich environment. Dope users only need to wait until they have metabolized the chemicals in their bodies. These are luxuries psychotic people do not have.

* * *

If we react the same way as other mammals to extreme deprivation of external stimuli, then why are we so different in normal circumstances? Why are animals, but not people, bound to an immutable repertoire of behaviors? Why are our minds so constantly, so restlessly searching for things to keep themselves busy, whereas those of animals are not? Why do we always need to be *doing* something—playing silly games, having endless trivial conversations, or staring at the television for hours on end? Why do we stubbornly keep exploring and experimenting all our lives, while most animal species only ever play during their youth, when they are acquiring and practicing their normal behaviors?

We know that our fellow mammals have an inner world of their own, so it cannot be the lack of an inner world that accounts for the difference between them and us. Instead, it must lie in the quality of these worlds. Obviously, the flexibility and versatility of the human inner world is a far cry from that of all other creatures. And let us not forget its sheer computing power. As we shall see, the reason for the gap between them and us is a combination of capabilities that spring from this powerful inner world.

Foremost among these is our sense of time. It is this sense that makes us aware of duration—but it can be deceived rather easily. Given enough random sensations in rapid succession, we are liable to start thinking time is flying. And there is a special kind of minimal activity that enables us to keep the dreaded beast called boredom at bay: watching something that moves. If you have ever tried just sitting on a chair in a silent room for half an hour, without speaking or singing, with absolutely nothing to do, you know that the boredom can be crushing. But strapped into the passenger seat of a car with an uncommunicative driver, we can ride for hours without complaining. The difference is the scenery whizzing by. It does not even have to be interesting or new. Millions of people travel the same route day in, day out for years until they know every bend, every tree, and every sign along the way. And still their boredom remains manageable, even with no radio or conversation.

The same goes for buses and trains, where we can sit and look out the window for hours on end. But this does not work on planes—there is just too little to see. The windows are small, and it is hard to make out much more than endless clouds, grey ocean, or a hazy, featureless landscape far below. Paradoxically, flying is much faster than any other mode of transport, but it often seems to take forever. We quickly get bored, restless, and irritable, which is why cabin crews come around so often. Their main concern is to keep you busy.

And there are other simple but remarkably effective ways of alleviating boredom. The ubiquitous aquarium of Chinese restaurants does wonders for the lone diner. The comforting effect of a crackling log fire is strong enough to work even if it is on a DVD. Television works in much the same way. And let us not forget that greatest pastime of them all: enjoying the view. Just sitting outside on a terrace somewhere, in a meadow or on a ledge, staring out across mountains and valleys, fields, and woods, or even traffic and passersby, is more than enough to keep most of us entertained.

There is something very peculiar about enjoying the view: it is an adults-only thing. Children the world over find it infinitely "booooo-ring!" This contrast may actually be the key to understanding why all that sitting around and staring into the distance is so conducive to our mental well-being. The act of taking in the view closely resembles one of the most important activities in the history of our and that of all other social species: keeping watch. Without a doubt, a willingness to keep an eye out for anything untoward in their surroundings was selectively advantageous to our ancestors, especially after they moved from the natural cover of the forests to the open savannah millions of years ago.

Those surroundings were never completely motionless. Grass quivers in the wind, leaves rustle, water flows and ripples and shimmers in the sunlight, clouds drift through the sky, and bees, flies, and other insects buzz by. If it is warm and sunny enough, everything in the distance seems to be gently vibrating. Anyone saddled with guard duty must have appreciated all this, not because of its inherent beauty, but because it made it easier to keep on keeping watch, constantly scanning the horizon for approaching danger. Essentially, evolution shaped us to enjoy sitting on a ledge, perched at the entrance to a cave or in the shadow of a tree, looking around for long periods of time even when there is nothing interesting or unusual to see. Today, the need to keep watch is largely a thing of the past—yet we still enjoy the soothing effect of playing sentry by drinking in the view.

Unlike adults, children are hopeless at keeping watch. They are sensitive to and interested in anything that moves, and they positively love having the window seat in cars and trains. But the sentry aspect is beyond them. Standing guard is a part of our genetic behavioral repertoire that does not flourish until later, at the end of puberty, and even then it shows up only in that odd remnant of days long gone: enjoying the view. Actually standing guard in the military sense is one of the most boring, least popular chores a soldier can be saddled with.

But these innate vestiges of ancient behaviors that stave off boredom do not explain why we get bored so much more quickly than animals, and why the human mind is so restless. Nor do they explain why all sorts of thoughts, feelings, and urges are constantly and spontaneously bubbling up from it. To understand all this, we need to take a closer look at the way our inner world works.

* * *

The spectacular growth of our forebears' brains gave them an enormous amount of extra brain capacity. First, this must have vastly improved their memories. In turn, this would have made learning from experience much easier than it was for species without such a wealth of cortical tissue. It enabled them to store memories of experiences and respond more quickly and effectively whenever similar situations arose—an obvious evolutionary advantage. It made hominids more versatile and more flexible, which enabled them to adapt better to changes in their environment, so much so that they managed to eke out an existence in almost any habitat, including unusually inhospitable ones. As a result, *Homo erectus* was able to spread successfully across all of Africa, Asia, and Europe on the strength of a brain just twice the size of that of his savannah-bound predecessors from a million years before.

And yet the gain was limited. A better memory will help you respond better to known threats and situations, but it does not help you dream up new ways of using those circumstances to your advantage. Reactions based solely on memory are purely behavioristic achievements; they are automatic responses to whatever external stimuli happen to present themselves. So there is no response, no action, without a stimulus. Such a lack of spontaneous activity is still very far from the characteristic behavior of modern humans.

Nevertheless, the increased storage capacity of our ancestors' memories brought with it at least a modicum of new possibilities. The growing wealth of memorized data may have led, in time, to the development of new "memories" not directly caused by external input. These may have been the first true products of the inner world itself. Memories, at the most basic physiological level, consist of an intricate network of neurons, the synapses of which keep each other in a state of excitation. If such semipermanent networks overlap, this might give rise to new patterns of activation. These patterns—let us call them secondary memories—have no direct relation to any external event, and yet for all intents and purposes are indistinguishable from true memories of past experience. The cleverest animal species show signs of this process too. But the unprecedented growth of the human brain began to open the door to much, much more.

Meanwhile, some capacities were developing inside the growing hominid brain that not only increased their chances of survival, but might also bring about more flexible management of actual and secondary memories. One of these was the superfast ballistic computer that enabled them to throw accurately. This neural computer is not just speedy. As anyone who is ever seen the mass of work involved in aiming and firing old-fashioned artillery will attest, it is also capable of handling vast amounts of data simultaneously. In today's terms, we might compare it to a top-notch gaming computer—and we would still be doing the brain an injustice.

Two other, equally important new capabilities also came to reside in our forebears' brains. One was the ability to sing. The other was a good sense of rhythm. To start with the latter, rhythm is an essential feature of any organism, from the highly complex all the way down to the single-celled. Every instance of controlled and coordinated movement, from the "swimming" of bacteria to the galloping of a racehorse, depends on rhythmic regulation. And the whole gamut of cyclical phenomena in the lives of organisms revolves around rhythms: patterns of sleeping and waking, the daily cycle of body temperature, the movement of food through the intestines, the functioning of the heart, the respiratory cycle, the reproductive cycle, and so on. All these are determined by a variety of biological and biochemical clocks located inside the body or even inside individual cells.

Natural rhythms tick at different speeds. The heart's rhythm is relatively fast, especially in small mammals and birds. In some species, it can reach up to five hundred beats per minute. The sleep cycle is diurnal, meaning it follows a daily rhythm. So too does that of body temperature. Human menstruation is a monthly cycle, which seems to derive from the moon cycle, as does the idea of the month itself. The week, by contrast, is a cyclical concept that is as hard to explain as it is resilient. All historical attempts to introduce more logical or more efficient calendars based on weeks shorter or longer than seven days have failed. Nobody knows why; in fact, we do not even have the vaguest notion of the thinking behind the seven-day period.

Luckily, we do not need to explain it. All that is important here is that rhythm provides structure and regularity. It is an ordering principle that enables us to process and manage large amounts of data, such as sensory stimuli and bodily sensations. With its help, we can manipulate all these data and group them into larger, meaningful complexes.

Bodily functions are not the only aspects of organisms that revolve around rhythm. It also plays a part in mental processes. We know this because group behavior, too, seems to be mediated by rhythm. Think of the mating behavior of many animals, with their intricate dancing and turn-taking rituals. Many large species of birds, for example, such as swans and crested grebes, put on impressive performances in the mating season.

But their dancing duos are not the only ones to show highly coordinated group behavior—apes do too. This may come as a surprise, given the classic cliché of chimps and gorillas leaping about wildly and aimlessly, thumping their torsos and roaring savagely at anything and everything. Proof that the cliché is wrong comes from that old film clip we came across in chapter 2, in which two young chimps try to reel in a crate of bananas. The box is out of reach, but there is a rope lying loosely around it. By each pulling one end of the rope, they can inch the bananas their way. But if they do not pull in time with each other and with equal strength, one of them will end up with nothing but rope and the other with nothing at all. As we saw earlier, this is not what happens. The pair set about their task with perfect rhythmic coordination until finally the bounty is theirs.

Rhythm is also a cornerstone of the human language faculty. For one thing, it can turn prose into poetry all by itself. The oldest Western literature, Homer's *Iliad* and *Odyssey*, is mainly in dactylic hexameter, lines of verse consisting of six feet of three syllables each, the first of which is stressed. Dactylic verse sounds like a waltz, like this, where the boldface indicates stress:

Six times a | **dac**tylus, | **each** of three | **syl**lables, | **make** a hex | **a**meter.

Such poetic forms were not only used for their aesthetic appeal. They also had a purpose. In Homer's day and long afterward, stories were spread and handed down over the generations by word of mouth. The narrator was usually an amateur or an itinerant professional bard who could no more read or write than his audience. Even if he had been able to write, this would have been of little use, for there was no practical and affordable medium for carrying information. Paper had yet to be invented, papyrus was unavailable outside Egypt, and parchment was far too costly. Lugging around huge stacks of clay tablets was hardly an option. So the tales, some of which were thousands of lines long, had to be learned by heart. This is where their metric regularity and the standardized length of the verses came in handy. So too did the literary stopgaps and clichés with which such stories are riddled. These range from set combinations of two or three words—in Homer's epics, dawn is always *rhododaktulos,* or "rose-fingered," and the Goddess Hera is without fail *bo-ópis,* or "cow-eyed"—all the way up to quite lengthy formulae. With the help of these tricks, the bards were able to remember the poems with a minimum of blood, sweat, and tears.

Our language faculty uses rhythm in lots of other, often very subtle ways, something we generally remain blissfully unaware of. For one thing, rhythm is an important technical tool for structuring words and their inflexional forms. It is instrumental in determining the stress patterns and melodies of sentences, including which sounds we can leave out and which linking sounds we need to insert, and where, to allow for fluent speech. If rhythm serves all those purposes, it is likely to be relevant to structuring thoughts and trains of thought as well.

Closely connected to rhythm are singing and all the rest of our musical abilities. But they are more than rhythm alone. Music also requires a sort of syntax, a system of principles and rules that determines how tones can be strung together into larger units, which intervals are allowed, and what is considered off-key. The details of musical grammars differ from culture to culture in ways reminiscent of the differences between human languages. And yet, musical abilities are not unique to mankind any more than rhythm is. Many types of small birds are great singers, and dolphins and other whales are known for their lengthy, intricate melodies.

The behavior of sea mammals in the wild is extremely hard to observe, which is why we know very little about how they spend their days. This also means we have not the foggiest idea why they sing at all, let alone what their songs are supposed to convey, if anything, and to whom. Birds are easier. Only male songbirds sing, and they do so for the age-old purpose of scoring girls and warding off the competition. For us humans, the ability to produce and enjoy musical patterns seems to have had evolutionary value in three ways.

First, men may have started singing for the same reasons as birds: to attract and seduce females. Whatever its exact origins, this certainly became an important function of music at some point. It is hardly a coincidence that love and betrayal have been the subject of songs for as long as anyone can remember. Throughout history, countless Romeos have won their Juliets through the power of a well-executed serenade. When the Beatles appeared on the scene in the early 1960s, girls would literally swoon in droves at the mere prospect of seeing them in the flesh. At one fabled concert at Shea Stadium in 1965, the screaming was so intense the fab four could not even hear themselves play. Later rock 'n' roll gods never wanted for female attention either. Countless are the stories about girls lining up outside artists' dressing rooms, desperate to get in. Curiously, screeching fans typically throw themselves at the feet of the singer, while the other band members get relatively less attention. Singing obviously has an exceptionally erotic charge.[1]

The second evolutionary advantage of singing for the human race is that it helps coordinate behavior. Both part-singing and antiphony, a call-and-response type of singing, are useful tools for groups of people engaged in collective exertion. They help us to maintain a steady pace, keep up morale, and get the most out of our muscles. When it comes to hard labor, heaving and hauling, ramming and pounding, cleaving and sawing, digging and sifting, everything goes better with a song. So does marching, as any good foot soldier knows. Singing ups the strength and effectiveness of almost any group of people involved in coordinated action.

Last but not least, singing of any kind helps form emotional ties between people. We have all felt the immediate bonding effects of singing "*Happy birthday*" or the national anthem. But one person singing to an audience works just as well. Think of the mass waving of lighters, or these days smartphones, at concerts and festivals. Such bonding events may have been more common before radio and television made entertainment a much more individual experience. But anyone who has ever sat around a camp fire with a few friends and a guitar knows the feeling.

* * *

So our ancestors had quite a number of tools for bringing order and structure to their ever-expanding inner worlds. They had the capacity for controlled throwing, which enabled them to process large amounts of data at lightning speed. They had a well-developed sense of rhythm as well as the capacity to sing. But it was one of these faculties—most likely singing—that led to the real breakthrough: recursion.

The nature of recursion is best explained by the peculiar behavior of ordinary rabbits. Essentially, the whole rabbit universe revolves around two processes, growth and multiplication. Let us call them the Law of Rabbitual Growth and the Law of Rabbitual Multiplication.

The Law of Rabbitual Growth states that if a rabbit is combined with a carrot, the result is more rabbit. According to the Law of Rabbitual Multiplication, however, if a rabbit is combined with another rabbit, the result is more rabbits. Not a larger individual rabbit, but a larger rabbit population. The important point is that such processes operate on their own output. In the case of multiplication, the output of the process is in fact the only thing needed as input: rabbits. Rabbits in, more rabbits out. This is recursion at its finest: a process that generates everything needed to make it run again, and again, and again.

Not many processes are like this. The Law of Rabbitual Growth is only weakly recursive. The process does generate some of the input

I. Law of Rabbitual Growth.

II. Law of Rabbitual Multiplication.

needed to make it run, namely "rabbit." But after one cycle, the other part of the required input—the carrot—has been consumed, and the process comes to a halt. Unless we go and fetch a new carrot from somewhere, the rabbit cannot eat and will not grow.

The world is full of processes that are not even weakly recursive. Take glass manufacturing. Much simplified, the process of making glass looks like this:

sand + salt + sodium carbonate + heat → glass

Here, the end product is of a kind quite unlike any of the elements needed to make the process run. So making glass is not a recursive process. Strongly recursive processes generate all their own fuel, so they can run on their own with no external interference. The results need not be absolutely identical to what was put in, as long as input and output are sufficiently alike to count as "of the same kind." Just as the Law of Rabbitual Multiplication yields ever more rabbits, all slightly different and in that sense slightly "new," all strongly recursive processes can go on generating new outputs of the same kind as their input indefinitely.

It is not all that difficult to see what a system containing recursive rules, like the rules of singing, might mean for the expanding brains of our ancestors. These recursive processes might be applied to inner world products, generating an endless stream of new, slightly different, somewhat more complex and more varied inner world products. It is this that underlies our continually restless mind, ever bubbling over with new ideas and thoughts. As a consequence, our minds are always hungry for impulses large and small to keep themselves busy. Only death or severe brain trauma will put a stop to the self-propelling whirligig in our heads.

Actually, our brains are filled with recursive rule systems. Arithmetic, for one. Regardless of the number system used, be it base 2, base 10, or any other, addition always boils down to combining two numbers to make another number. In other words, the process uses nothing but numbers to produce other numbers—a perfect example of recursion. Another such system is language. The recursive nature of the rules by which we construct words out of smaller words, and sentences out of smaller units, is what enables us to produce an infinite array of new expressions. In the most literal sense, we can never run out of things to say.

But how did all this machinery develop? At the very least, it required the integration of our inner world and our throwing skills, rhythm, and singing capacity. Only when these began to merge did our minds reach a level of sophistication and abstraction capable of dealing with complex inner world products. And only *then* were our brains able to create a self-made version of ourselves, an image divorced from our physical reality and our environment—a purely psychological "I," on a par with all the other notions and concepts that populate our minds. Only then did we become true individuals. As a consequence, only then did we develop an image of and about ourselves: a soul. A soul which has its biochemical roots in the brain, but is immaterial and spiritual (although not at all supernatural). It remains just as bound to our brain as any other thought or mental sensation, and it will die with us.

* * *

At the same time that the integration of this whole set of specific mental functions made us masters of our own inner worlds and paved the way for abstract concepts like "I" to arise, it laid the foundations for the emergence of the human language faculty. In fact, they are two sides of the same coin. The mechanisms we use to produce language are the same mechanisms that enable us to order and structure our inner world. And inner world processes like conceptualizing and categorizing sensations are essential parts of the language faculty as well. The same is true of recursion and the kind of rule systems we call syntax; these are at the heart of the language faculty, but they are also fundamental to singing and music. Rhythm is a defining property of our biological existence, and at the same time a cornerstone of our language faculty. The machinery involved in high-speed ballistic computation, needed for controlled throwing, is the same machinery that enables us to produce more than 150 words per minute. And our urge to point and name, which also seems to derive from our throwing ability, is one and the same driving force behind the acquisition of vocabulary.

This also explains how we go about constructing highly structured, complex trains of thought. We do our "serious thinking" and reasoning in a highly conscious way that relies on language. Thinking and reasoning are not unlike doing arithmetic. Our automatic mental pilot can take care of simple responses to fairly standard stimuli—the things we do "without thinking"—just as it can handle small numbers without recourse to formal reckoning. We may even give our inner world free reign for a while, away from the prying eyes of consciousness. This is what we do when we say we want to "sleep on it," which is a good way of mulling subconsciously over problems that primarily involve moral or personal choices. But when it comes to things that are more complex and formal, we start thinking more consciously, monitoring those thoughts, and formulating them by way of words and sentences. We may even resort to actually making notes in order to keep track of what we are thinking.

The idea that integration of already existing brain functions gave rise to both conscious thinking and language also makes good evolutionary sense. It solves the three great mysteries that have always beset every attempt to explain how the language faculty came about. The first is how something as complex as the language faculty could have developed without advance planning. The second is how the enormous amount of energy spent on building a system that only became operational after hundreds of millennia of construction time could be justified. The third is what the evolutionary advantage was at every stage of development.

The key to all three is in the fact that each of the systems and functions involved developed for its own reasons, their growth and continued improvement promoted by a clear evolutionary advantage to the individual at all times. Throwing just a little bit faster and more accurate, being just a tad better at coordinating your efforts with others, cooperating with others, singing your way into the hearts of the girls just a bit more effectively, all enhance your chances of surviving and passing on your genes to a new generation. So no advance planning is needed. By the same token, no energy was spent on constructing a system that will only become functional sometime in the future. Every calorie spent on any of the brain functions involved was immediately balanced by an increase in evolutionary prowess.

Only when all these systems were in place could the integration process have begun. The extra cost of integrating existing systems would be negligible compared to the gains in terms of an ever more powerful, more intelligent brain. Integration also solves the mystery of why our forebears showed virtually no sign of intellectual activity for so long after their brains had reached their current size and their bodies had become fully modern. Integration only began when our ancestors had become anatomically indistinguishable from ourselves, and it took time to reach a level of integration that generated the individual, detached "I" and the language needed to manage and enhance abstract thinking—and with them our insatiable thirst for innovation.

Best of all, the integration hypothesis quite elegantly takes the communication problem off the table. Remember how the idea of communication as a driving force flew in the face of normal Darwinistic reasoning? Remember how, if communication was central, language would be only advantageous to you if *others around you* had it, which is Darwinistic gobbledigook? Under the integration hypothesis, language evolved primarily as a means for creating order and housekeeping in our own inner world. Its development allowed for more sophisticated thoughts, made us more intelligent, which is a real evolutionary advantage. For some reason, the budding language faculty must have found a way out into the world in the form of speech. Perhaps speech began as a self-monitoring device, serving roughly the same purpose as taking notes on paper when we are really thinking hard about something important.

That said, communication has obviously become enormously important—but only after the fact. Just like integration began only when all the systems involved were in place and fully functional, linguistic communication began only when integration had progressed far enough for language to have emerged. We might view language and thinking as exaptations of the systems underlying them, and communication as an exaptation of language. Its importance can hardly be overestimated, for the communicative qualities of language released our forebears from the isolation of their own minds. Now that they could express their thoughts, they were able to share them with other people, and at the same time understand what other people were thinking. All of a sudden, they could pool their intellectual resources. They could put their brains together, so to speak, and benefit from the power of collective thought. And all this came at no extra cost. All it took was the willingness to hear each other out.

* * *

The idea of the integration of mental faculties per se is not a new one. In 1976, American psychologist Julian Jaynes offered an explanation for the emergence of consciousness based on integration. Twenty years later Stephen Mithen, a British archaeologist, went down the same path in search of the origins of culture and art.

Jaynes saw the brain as divided into separate compartments, which remained closed to each other until about 1000 BCE, almost Homer's time. For Mithen, however, integration took place some two million years ago, around the time when hominids first began to make stone tools.

In Jaynes's view, consciousness is a byproduct of human language, so it arose after and as a result of language. But dating this mental revolution so late raises the question of how consciousness could have evolved roughly simultaneously everywhere on earth, across groups of people who had been living in isolation from one another for tens of thousands of years. If this theory were true, why did nobody ever come across a tribe of preconscious people, not in long uncharted Australia, not in the remotest valleys of New Guinea, nor even in the deepest recesses of the Amazon jungle?

Mithen, on the other hand, falls victim to the common fallacy of mixing up language and speech, and so invalidates his own theory from the outset. But there are more problems. In Mithen's view, the language faculty develops in two stages, the first taking place around two million years ago, the second coming to completion somewhere between five hundred thousand and two hundred thousand years ago. What he cannot explain is why people who possessed a full-fledged language faculty then completely ignored it for several hundred millennia, only to embark quite suddenly on a veritable spree of cultural and technical innovation some ten thousand years ago. Mithen's theory is not only improbable but also utterly untenable from the perspective of evolutionary cost.

None of these problems need bother us if we see integration as a process that began recently enough that all the systems involved were fully developed and operational, and long ago enough to explain its ubiquity. A process that began slowly, on a small scale, and gradually picked up pace until it reached a point where its presence became apparent in the form of our language faculty. A process that may also have had other consequences for the way our mind works—and that may in fact still be ongoing. We might look upon it as a slow trajectory of increasing enlightenment, punctuated by the odd landmark change when certain things came together, little by little enhancing our understanding of the world we live in and our position in it. It was a development that not only gave rise to our individuality but also to human language and, further along the same road, ever new forms of rationality and logical and mathematical insight.

So when did all this kick off? Naturally, pinpointing the starting date of this process is tricky. But we can find a few clues lurking in the field of paleontology. First, although the hominids of the Lower Paleolithic rarely seem to have come up with anything new, *Homo erectus* was clever enough to manufacture and use flint tools in great number. This is a clear sign that they could act not only purposefully, in line with a plan, but also methodically, following an established technical procedure. The other achievement attributed to *Homo erectus* is the taming of fire. But here the picture is a lot murkier. It may well be that, rather than truly mastering the art of lighting things up, they made do with fires that were accidently sparked by lightning or by their own flint pounding as they made their tools. This would have required much less insight and planning than keeping their fires burning permanently, say, or even transporting them in the form of smoldering dry moss. We have no way of knowing what went on in that strange, long-lost world. All we can say for sure is that the hominids' intellectual achievements were far too poor to suggest integration was already underway in their day.

The appearance on the scene of anatomically modern humans, who some two hundred thousand years ago began to replace *Homo erectus* and ultimately also the Neanderthals, might seem like a natural starting point for integration: if they had our bodies, then why not our brains? The problem is, we have no way of knowing whether all the systems involved in integration were already in place at that time. And no evidence of any kind of integration dating from that period has ever been found. As far as the paleontological record goes, early *Homo sapiens* lived much like their hominid ancestors. So it could well be that integration did not begin until much later.

The *latest* date by which integration must have started is when the ancestors of the Aborigines colonized Australia. They were cut off from the rest of mankind until Dutch explorer Willem Janszoon first sighted the continent dozens of millennia later, in 1606, and the occasional European ship began dropping in on the Australian shores. Still, contact with the inhabitants remained minimal until 1770, when the continent was formally claimed for Britain by Captain James Cook. But when and how Australia was first colonized is unclear, to say the least. The oldest human fossils on the continent are the remains of what seems to be a man buried at Lake Mungo, about three hundred miles north west of Melbourne. Discovered in 1974, his demise has been dated to anywhere between twenty-five thousand and no less than sixty thousand years ago, depending on who did the measuring and what methods they used.

The problem with Australia is that it has always been completely surrounded by sea, even at the apogee of the ice ages when the global sea level was at its lowest. So, as the standard reasoning goes, people probably reached Australia around forty thousand years ago, when the gap between it and the Indonesian island of Timor was at its narrowest. This tallies with the fact that *Homo sapiens* was quick to spread all over the rest of the inhabitable world, and with most estimates of the age of Mungo Man, as he is known. Specialists still disagree as to exactly how those first settlers reached Australia, but most assume they must have been highly skilled seaman and shipbuilders. This, in turn, implies a level of planning and abstraction that is inconceivable in the absence of a truly modern mind and a fully operational language faculty. If this is the case, integration must have started much earlier still. However, there is a lot that makes this kind of account hard to swallow.

First of all, early *Homo sapiens* conquered the world, including many regions that are now islands, by foot. This was possible, thanks to the low sea level during the ice ages, which meant most regions were connected to the Afro-Eurasian landmass by land bridges. Only the Australian mainland, Oceania, and the Americas were not. And of these, only Australia was reached at a relatively early stage. North America was not colonized until twenty thousand years ago, when its first pioneers made their way overland from Siberia. The many tiny islands of Oceania had to wait until Christian times, long after really seaworthy boats and competent sailors had arrived on the scene. Other than the undeniable fact that people had reached Australia, there is no indication that boating or rafting of any kind existed before five to ten thousand years ago.

If we do accept, nonetheless, that the original colonists of Australia were intrepid seafarers who ventured out on purpose across the Timor Sea, then we too find ourselves up a creek without an oar. What are the chances that a people who had achieved such unique maritime prowess so early on would promptly lose all interest in boats and sailing forever? Where are the memories? Why is there no mention of such a momentous event—or of any seafaring at all, for that matter—in Aboriginal mythology? And why is there no evidence of early boating anywhere in the Indonesian archipelago? It seems more likely that the first people to reach Australia did so by accident. A group of survivors of some natural disaster, a great storm or a flood perhaps, who drifted across clinging to an uprooted tree or a tangle of driftwood.

If Mungo Man's ancestors washed up on Australian shores by accident, they might have arrived before the integration process was complete and may not have had language at all. But as it turns out, the value of Mungo Man in determining the starting date of integration is doubtful in any case. Around the turn of the twenty-first century, a graduate student named Greg Adcock rather ingeniously managed to extract some DNA from Mungo Man. His results, which are not entirely undisputed, indicated that Mungo Man was not genetically related to modern Aborigines. What is more, at six feet, five inches Mungo Man was much taller and lankier than the average modern Aborigine. This suggests that, in the long run, Mungo Man's people were unable to cope and eventually died out.

So Australia must have been colonized at least twice. Since Mungo Man's people do not seem to have mingled with the ancestors of modern Aborigines, this second colonization may have happened much, much later—perhaps at a time when sailing had become more common and our language faculty was already up and running.

* * *

So the early Australians did not help us tie up all the loose ends. Yet, there are good reasons to assume that the integration of our brain functions was well underway by forty thousand to thirty-five thousand years ago. This is around the time the Cro-Magnons produced their first cave paintings. As we have seen, these paintings seem to testify to a kind of wonderment about certain things the Cro-Magnons came across in their world, a kind of vaguely conscious experience of them. In this way, they send a signal across the ocean of time that our ancestors were making headway on the journey towards the dawn of individual consciousness.

But in over twenty-five thousand years or more of cave painting, right up until they vanished mysteriously when the land ice finally retreated, the Cro-Magnons never reached the point at which they immortalized themselves in their works. With one or two exceptions, the only things heralding the emergence of the first rudiments of a notion of self and consciousness are the tentative signatures they left in the form of handprints and patterns of dots.

This is not to say they made no images of humans at all. In fact, they did, all over Europe and on a fairly large scale. But these images exclusively took the form of small, portable figurines. Best known among these are the so-called Venuses.

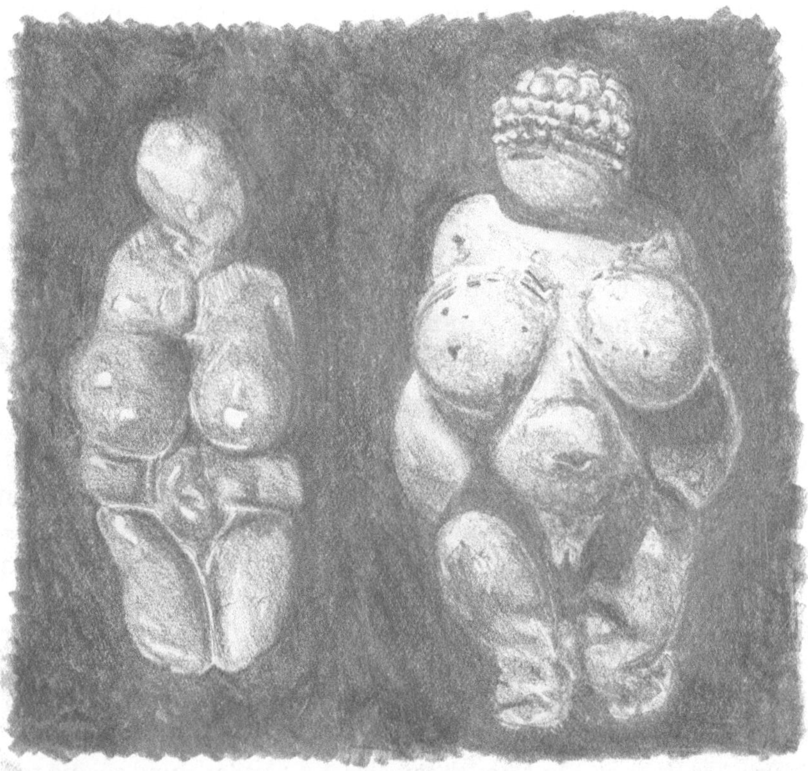

Some typical Venuses, with the Willendorf Venus on the right.

The mother of all Venuses is without a doubt the Willendorf Venus, a limestone statuette discovered in Austria in 1908. At 4.5 inches tall, it depicts a bloated woman with a meticulously sculpted coiffure, in the kind of neat little curls also frequently found in classical sculpture. But the Willendorf Venus is no less than twenty-five thousand years old. And she has no face.

This is no accident. No statuette or figurine from Cro-Magnon times has a face. At over thirty thousand years old, Fanny, dubbed the ballerina of Galgenberg, predates even the Willendorf Venus by some five millennia. Found in 1988 at a site not far from where the Willendorf Venus had been unearthed a lifetime before, she is the oldest female figurine ever discovered in Europe. Her elegant pose inspired her rescuers to name her after Fanny Elssler, the nineteenth-century Austrian ballerina. Her impressive age notwithstanding, she is a lively, even lustful girl, deftly cut out of a shard of slate three inches long and no more than an eighth of an inch thick. She has one hand on her hip, the other at her neck, elbow turned out, almost as if she is leaning against a doorpost. As primitive as his world was, this craftsman knew exactly what he was doing. To complement the right breast, shown in profile, he even managed to suggest a protruding left breast using a miniature bas-relief. Ultimately, Fanny is the spitting image of a modern pin-up girl, radiating life and glamour. But she, too, has no face.

Faces on figurines do not show up until roughly fourteen thousand years ago, around the same period that images of people began showing up on the walls of caves. And they were first seen in the Levant, the area east of the Mediterranean. Around that time, another new kind of art was born: images and carvings that are unabashedly sexually charged, erotica almost. There are figurines and statuettes of couples making out and copulating, some of them in shapes that suggest they were expressly intended as sexual toys. The contrast between these and the older Venuses could not be more striking. Although the Venuses are nude, they have no noticeable sexual connotation; instead, they seem to point to motherhood and fertility, or sometimes to menstruation.

The Cro-Magnon miniatures share another striking characteristic: virtually all represent women. This might be an indication that sculpting was a male thing, just as men would later on dominate the visual arts until at least the late nineteenth century. It also ties in with the idea that the cave artists were able to conceive of and portray the most spectacular animals they saw, but not themselves. A similar thing seems to be going on with the male sculptors. To them, women were highly salient

"Fanny," or the Venus of Galgenberg.

beings. And so they managed to pick women out from the general background of their perceptual experience, but not their fellow men. They would not be able to see themselves and other members of the "unmarked" sex, as they saw it, until their budding "I" was developed enough to make them true individuals with clear, separate identities.

* * *

So all evidence on the mental condition of our early human ancestors points in the same direction: before roughly fourteen thousand years ago, virtually no images of men were produced. The cave paintings and surviving figurines—invariably carved from bone or stone, as the wooden artifacts that were surely also produced have long since rotted away—almost always represent either animals or women. And unlike the animals, the women never have faces. Images of men are conspicuously absent, as are erotic overtones.

All that changed thirteen to fourteen thousand years ago, first in the Levant area, in what is known as the Natufian Revolution. This is when the first pornographic imagery was produced, along with the first dildos and dildo-like figurines. And this is when figurines suddenly got faces. This is a clear sign of a sense of individuality and conscious personal experience on the part of the sculptor. Here, for the first time, we encounter people who must have been able to really "see" the other as an individual distinct from the general environment. A people who were able to "see" themselves—and who felt the urge to express it.

There is one major exception to this general picture. Over twenty-five thousand years ago, events took place on the inhospitable glacial prairies of Central Europe that would not occur anywhere else until more than ten thousand years later. Around 1920, near a place called Dolni Vestonice in the Czech Republic, the remains of a hunter-gatherer settlement were unearthed in which people not only produced ceramics in ovens but also the odd figurine representing a man. Later, similar finds were reported from nearby Pavlov and Brno. Strangely, the earthenware these mysterious early humans made was not of a sensible, practical type, like pots and pans, but mainly decorative, fired-clay figurines of animals and women. Still more surprising were the imprints found on lumps of nearby sediment, which could only have come from fabric made on some kind of loom. Looms, however, would not be invented until much, much later. These finds indicate a degree of cultural sophistication that would seem unattainable for people not in possession of a fully functional language faculty.

Another curious case is a settlement discovered at Sunghir in the Volga basin in Russia. At the time it was inhabited, some twenty-three thousand years ago, it was perched no more than ninety miles from the edge of the glacial ice. Here graves were found that, for their age, were filled with quite startling riches. One contained a man who must have been about sixty, together with almost three thousand ivory beads. The beads were probably part of a wealth of necklaces or other ornaments adorning his body or clothes when he was buried. He seems to have been wearing a hat, too, embellished with the teeth of arctic foxes. On his arms were no less than twenty-five thin bracelets made of mammoth ivory.

Even more surprising was the grave of a child, aged around thirteen. Some five thousand beads were found with the boy, far more even than those buried with the nearby man, although the beads themselves were smaller. Also with him was a cap with more than 250 arctic fox teeth, in addition to a number of other ornaments and an eight-foot spear made of mammoth ivory. It was a beautiful but completely impractical object, if only because of its weight.

But that was not all. The pièce de résistance of the site was the grave of yet another, smaller child, probably around eight years of age, draped with a total of 5,274 beads. This child was probably a girl, judging from the fact that her cap was embellished with beads, but no fox teeth. She was buried with a couple of toy spears made of ivory, among other things.

These graves demonstrate two things. First, the people at Sunghir took great pains when it came to burials. Second, they had a well-defined social hierarchy based on birth rather than personal merit. Each bead must have taken a good forty-five minutes to make, meaning that at least thirty-five hundred hours—slightly over two years' work—were spent on the garb of the two children alone. What with all the other riches in the graves, these would have been extremely costly funerals. In today's terms, only the very rich and famous would be able to afford such ceremonies. And if the Sunghir people had their own rich and famous, so to speak, they must have had an articulate system of social organization and stratification, unparalleled by anything else for many thousands of years.

Perhaps these inhabitants of what is now Central Europe and Russia were way ahead of their time. Perhaps Dolni Vestonice and Sunghir are where the first voices sounded out real words over the fields. Perhaps the brain functions of these people became integrated far, far earlier than they were anywhere else. If so, there is a tough lesson to be learned from their disappearance: no evolutionary advantage is by itself enough to save us from the vicissitudes of nature. If they had not succumbed already, the lives and culture of these prematurely supermodern inhabitants of Dolni Vestonice, Pavlov, Brno, Sunghir, and who knows what other, similar settlements were surely snuffed out by the last great surge of the ice. It would take many more millennia for mental integration, and through it language, true individuality, and conscious identity, to become so widespread that no natural disaster could put a stop to it. Once that happened, the world would never be the same again.

Note

1. Unlike birdsong, human singing is not restricted to males. And female singers do have an effect on men, although not in the same way male singers excite fans of the fairer sex. Male admirers of female performers, from Zarah Leander and Dolly Parton to Joss Stone or Lady Gaga, never scream their lungs out. They do not swoon, throw themselves at their idol, or try to force their way into her dressing room. During the nineteenth century, the image of the smitten beau sending flowers and suggestive invitations to his favorite operetta singer or vaudeville artist was popular. But this is a far cry from the behavior of female fans, which seems to be steeped in the sort of violence reminiscent of the Dionysia festival of ancient Greece, where bands of women went on drunken, erotic rampages that often ended in raping or killing any man they got their hands on.

   Women on stage exude power. They impress men, even if their voices are as sweet and apparently innocent as that of Marilyn Monroe or Carla Bruni. Offstage, things are different. There is hardly a stronger cliché of feminine vulnerability with mildly erotic overtones than the lone peasant girl, singing sweetly as she makes her way through the meadows, the archetypical Little Red Riding Hood. Just take a look at the cinematic history, from old-fashioned pornography to Julie Andrews in *The Sound of Music.*

7

Dawn

"Grantchester Meadows" is by far the most beautiful reverie Roger Waters, front man of the legendary band Pink Floyd, ever wrote:

> In the lazy water meadow I lay me down.
> All around me golden sunflakes settle on the ground.
> [. . .]
> See the splashing of the kingfisher flashing to the water.
> And a river of green is sliding unseen beneath the trees,
> laughing as it passes through the endless summer making for the sea

With its soft guitar, a warbling skylark, and insects buzzing lazily in the background, it is a peaceful number. But—if you are sensitive that way at least—it also seems to hint at a sense of promise.

These meadows that so inspired Waters are not the stuff of fiction. Perched on the Cam River near Grantchester, they lie just southwest of Waters's native Cambridge. And they have a history. In the early summer of 1935, some eight years before Waters was born, a lanky young mathematician had lain there daydreaming too. His name was Alan Turing, and that day, among the rustling of leaves and the humming of bumblebees, he was struck by a flash of insight. From this insight ideas would grow that have changed the world even more fundamentally than the catastrophe of World War II, which was already looming on the horizon.

Turing was on the hunt for an answer to one of the toughest mathematical problems of his day, a question that the great David Hilbert had raised over three decades earlier. That question was whether there might exist a single, infallible method to decide whether a mathematical proposition was sound or not. A method that could prove the truth or falsehood of any conjecture, without exception, from the simple $2 + 2 = 5$ right up to the proposition that no natural number greater than 2 can replace n in $x^n + y^n = z^n$. Among the reeds at Grantchester, Turing suddenly realized that a method for making such decisions, if it existed, would have to be uncompromisingly mechanistic. It would be a system of reasoning that relied on the blind application of rules as cold, hard, and clear as the steel rods and pistons of a machine. Like a machine, it would work its way through a finite series of steps until it reached an unambiguous, definitive conclusion—because after all, a conclusion can only be trusted if each and every step taken to reach it can be checked and evaluated.

The imaginary machine Turing devised would become legendary. But even more important than the Turing machine itself, that afternoon Turing unwittingly laid the foundation for something nobody could even conceive of at the time: the computer.

Ever since *Homo habilis* began splitting flint two million years ago or more, all tools, appliances, and machines have served a specific purpose. You can do one thing with them, sometimes two or three, but that is usually it. A knife is useful for cutting, slicing, and stabbing things, but not for fetching water. You can drive nails into wood with a hammer, and screws too (if you are American), but you cannot use it to unscrew those screws. A fifty-yard tunnel drill is good for nothing but drilling a hole in the ground. And a Bulgarian Wonder Slicer is great for cutting up vegetables, or so they say, but useless when it comes to frying an egg.

Turing's machine was fundamentally different. It could not slice and dice, nor drill or fry. Its very point was that it had no single, designated use. Instead it could perform any task imaginable, as long as that task could be formulated using symbols arranged according to explicit, well-defined rules. For instance, it would be able to play chess. The pawns, bishops, and knights would be the symbols, and they would move around the board in strict accordance with the rules of the game. Eventually a pattern would emerge that we call checkmate and the game would be over.

Proving mathematical propositions is not unlike playing chess, or following the step-by-step instructions for assembling an IKEA closet. Providing the task boiled down to following a series of logical instructions and making logical choices about which step to take next, it was right up the alley of the Turing machine.

Such a machine consisted of three imaginary parts. Instead of a chessboard to move symbols around on, there was an infinitely long tape divided into squares, just as a reel of film is divided into frames. The tape could move to and fro under the second part of the machine, a "magic eye" that could see what symbol was in the square beneath it, if any. The eye could then erase the symbol in that square, replace it with some other symbol, or do nothing at all. Which of these it did depended on the third part: a table of the rules of the game, prescribing what was to be done at each point. The rules were of the sort, "If there is a dot in the square, replace it with a cross, and move the tape left. Otherwise, move the tape right."

As long as a Turing machine remained bound to one particular table of rules, it was not fundamentally different from other machines. Like them, it could only perform one task—running the particular program represented by its table of rules. But Turing came up with a brilliant innovation. He saw that the rules themselves were patterns of symbols too, patterns that could be placed on the infinite tape just like any other pattern. So in theory, the tape of a single machine could contain not only all the data the programs were to manipulate but also all possible programs themselves. A Turing machine with a tape like this would be a truly universal machine, capable of performing every imaginable task, because it could run every imaginable program—and this is the very core of the concept of the modern computer.

The idea was there, but a number of other factors were instrumental in turning Turing's theoretical Universal Machine into reality. One of these was the wave of technological innovation brought about by World War II, when every effort was made and every last penny spent to stay one step ahead of the enemy. In 1943, just eight years after Turing's afternoon musings in the meadows, Colossus was born at Bletchley Park in England. This was a top-secret code-breaking machine, far more powerful than any number cruncher the world had ever seen. Colossus only vaguely resembled what we now consider to be a computer, but not three years later the Americans would come a lot closer with the Electronic Numerical Integrator And Computer (ENIAC).

ENIAC, a thirty-ton behemoth with no fewer than 19,000 radio tubes, was an electronic device for calculating the trajectories of grenades and ballistic missiles. But although ENIAC was a huge improvement over Colossus, it still was not a true computer, because for each new task the tubes had to be reconfigured by hand. It was a never-ending refurbishing project. Time and again, it had to be manually turned into a different machine, fit for one task only. Real computers are not like that. They remain physically unchanged as they go from one task to the next. We just feed them different software, and what they do depends on the software we let them run, or, in Turing machine terms, which table of rules on their tape we instruct them to use.

Another factor was a man who was in many respects the antithesis of the geeky, introverted Turing. It was a loud-mouthed,

Four scientists holding approximately equivalent parts of the first few computers in existence. On the left is Patsy Simmers, holding a large component from the first computer, ENIAC. (US Army photo).

womanizing, exuberant Hungarian American who, when he felt like it, answered to the name of John von Neumann. Shortly after the end of the war, Von Neumann came up with a basic design for computers that would be fundamental to every programmable electronic device, from central heating thermostats to supercomputers, to this very day. According to Von Neumann's blueprint, both the programs that determine what a computer is to do and the data on which these programs are unleashed, are stored in the same permanent storage device, called a memory, comparable to Turing's tape. In von Neumann's day, this was an electromagnetic tape, which was later replaced by hard disks, floppy disks, flash drives, and ultimately "the cloud." From there, the relevant program and data are transferred to a central processing unit (CPU), the embodiment of Turing's magic eye. The program's instructions are then applied to the data one by one with the help of a relatively small working memory. When there are no more rules left to apply, the result is passed back to the permanent storage device.

Within a decade or so, the stage for the computer revolution had been set. But for the computer to become reality, the world now had to wait for two practical inventions: transistors and integrated circuits. It did not have to wait for long.

First, in 1947, the transistor was invented. That kicked off the miniaturization of electronics, making it possible to stash ever-greater processing power in ever-smaller boxes. It was an essential breakthrough, because the traditional radio valves were simply too large, too vulnerable, and generated far too much heat ever to allow for the construction of a reliable computer.

Then, in 1958, two Americans independently jumped the last hurdle together. Jack Kilby and Robert Noyce discovered that not only the fundamental parts of electronic switches, such as transistors, resistors, and capacitors, could all be made from silicon, but so too could the circuitry needed to connect them in order to form a useful, functional unit. In other words, a complete electronic circuit could be built on a single scrap of silicon. An entire radio in one piece, as it were, and infinitely smaller than the tiniest receiver ever to have been soldered together in the traditional way.

And so it was that only a quarter of a century after Turing's almost whimsical musings in Grantchester Meadows, the star of the modern computer was beginning to rise. At first, computers were exorbitantly expensive, handmade beasts filling entire building floors. But as fast as prices and sizes have shrunk, so too has the number of people around the world today whose lives have not been touched by computers.

* * *

In the summer of 1886, almost exactly half a century before Turing's epiphany, the people of the German city of Mannheim stood staring in amazement at a three-wheeled coach that coughed and wheezed its way through the streets, entirely unaided by a horse. The little vehicle puttered along erratically, reeling and bumping, and every now and again scuffing up against a tree or a garden wall, to the great merriment of the locals. Who had ever seen such a thing!

On top of the unruly contraption sat one Karl Benz, sweating profusely, bowler askew, trying valiantly to keep control of his invention. Benz was a local machine manufacturer and a pioneer of the internal combustion engine, which had developed from the gas engine invented by the Belgian engineer Étienne Lenoir in 1860. The curious mechanical capers that so underwhelmed the people of Mannheim that day were in fact a milestone in human history: what they were witnessing was the world's very first run of a motorcar with an internal combustion engine. The *Motorwagen,* as Benz had christened his horseless carriage, was powered by a four-stroke, water-cooled engine, a setup that has remained pretty much standard to this day. And Benz came up with many other components and designs that cars have had ever since, such as spark plugs and the throttling mechanism, carburetor, gearbox, and water-filled radiator.

That same year, another German pioneer of the combustion engine was building the first motorcycle. This was Gottlieb Daimler of Stuttgart, sixty miles down the road. Within four decades, their companies would merge into the industrial giant Daimler-Benz. Yet at the time, neither man knew of the other's existence. Nor had they ever heard of a clever Scotsman called John Boyd Dunlop. A veterinarian, Dunlop traveled the rough country roads of the day by carriage, dreaming of a more comfortable mode of transport. In 1887, acting on a hunch, he replaced the massive rubber tires of his young son's tricycle with hollow rubber tubing pumped full of air, and behold: the boy rode like a prince floating along on cushions. Dunlop applied for a patent and discovered that a previous, now-expired patent had been issued for a similar invention in 1845, to his compatriot Robert Thomson—which may say something about the appalling condition of nineteenth-century Scottish infrastructure. But Thomson had been too far ahead of his time. The technology for gluing rubber was not around in his day, and so it was Dunlop who ultimately took the credit for inventing the inflatable tire.

Next to the trailblazing work of Benz, Daimler, and Dunlop, people were busily experimenting with cars powered by steam and electricity. The first electrical car to be commercially produced, an invention by the French coach manufacturer Charles Jeantaud, dates back to 1881. These were heady times when it came to automotive engineering.

Things were moving fast in other areas as well. In 1888, not two years after her husband took his first drive, Benz's wife and their two young sons climbed into one of his little cars and went to visit her mother in Pforzheim, a sixty-six-mile trip fraught with problems. Where to get fuel? Who could repair the broken driving chain? But she got there, and her eventual arrival marked the successful completion of the first-ever long road trip. Five years later, Benz invented rack-and-pinion steering, a great step forward, and in the same year the French count Jules-Albert de Dion and his business partner, Georges Bouton, patented the design for the De Dion rear axle. This assembly, consisting of two independently moving parts, is still considered the best rear axle design ever made.

1895: Benz's daughters, Klara and Thilde, on an outing near Mannheim in a Benz Patent-Motorwagen "Velo," Hot on their heels is their brother, Richard, in uniform, driving a Benz "Phaeton." (photo Mercedes-Benz).

In fewer than ten years, the key components had been invented of every car that would be made until a full century later, after the advent of electronics. The rest, as they say, is history. The year 1894 saw the first endurance race from Paris to Rouen, in northern France. Although a steam car averaging an impressive thirty-seven miles per hour crossed the finish line first, a gasoline-fueled car was declared the winner. Five years later, one Camille Jenatzky from Belgium broke the magical one-hundred-kilometer barrier in an electrical, bullet-shaped racing car. Around that time the first hybrids took to the streets: cars with both a combustion engine and a battery-powered electrical engine. In 1903, the Dutch Spyker company produced a sixty horsepower beast, not only the first car in history to be powered by a six-cylinder engine but also the first four-wheel drive. And so the stage was set for the 1907 rally from Peking to Paris, a ten-thousand-mile monster race across unpaved, often virtually nonexistent roads. Five teams entered the race, four of which managed to finish it. The winner, after a solid sixty days, was the dashing Italian prince Scipione Borghese, driving an Itala. On reaching Moscow, he even found time for a flying detour to Saint Petersburg and back to attend a dinner party in honor of the event.

A year after this mother of all rallies took place, Henry Ford's factories in the United States began producing the famous Ford Model T, the first car to be manufactured by means of a novel, strikingly efficient production system: the assembly line. This was the final step in the maturation process of the motorcar from experimental toy for the adventurous few to mass commodity. From that moment on, cars would steadily become faster, more powerful, more comfortable, and safer—but there would be no more fundamental changes to the basic design of cars until the end of the twentieth century, when the advent of the computer and environmental concerns finally changed the game.

The entire trajectory, from experimental thingamajig to industrial, mass-market commodity, took less than a quarter of a century. Fifty years later, cars had become a fixture of modern life, traffic jams had become a familiar sight, and warfare was fully mechanized. Like computers, cars have drastically altered the lives of almost everyone on the planet.

* * *

There is a recurrent pattern to the kind of revolutionary invention that fundamentally changes the world: as soon as the idea has emerged and the technology becomes available, all the essential components and design features of the mature product are developed with astonishing speed. Both cars and computers went from the "wouldn't it be nice if . . ." stage to being practical and powerful tools in as little as twenty-five years. Once this stage has been reached, the invention spreads rapidly to all corners of society—both geographically and across socioeconomic classes. From then on, there will be endless minor additions and improvements, but the basic design will remain unchanged for a long, long time.

Steam trains and railways are another excellent example. The first breakthrough came in 1789, the year of the French Revolution, with the invention of the flanged wheel by the Englishman William Jessup. Jessup's flange made it impossible for carts, especially in the mining industry, to run off the iron plates that had long been in use as rails, into the mud. Within fifteen years, Richard Trevithick, a Cornish mining engineer, had mounted one of his high-pressure steam engines on a set of Jessup's wheels and was using it to pull around carts of ore and men at a Welsh ironworks. That was in 1804, and by 1829 Robert Louis Stephenson had turned Trevithick's rudimentary pulling machine into the first full-fledged steam locomotive, the famous *Rocket*. In all, it took a mere forty years to get from the flanged wheel to the Rocket, and steam locomotives have followed the design principles of the Rocket ever since.

Even faster was the emergence of public railway networks. It all began when a company built a railway track from Stockton to Darlington in northeast England in 1825, four years before Stephenson came up with his archetypal Rocket. In the meanwhile, it operated a regular service for passengers and goods using Stephenson's earlier, experimental engines. In 1830, the United States saw its first steam-powered railway line, a twelve-mile stretch of the Baltimore-Ohio railroad. One after the other, the nations of the Western world followed suit. The speed with which tracks were laid was positively intimidating. When the American Civil War broke out in 1861, the country already boasted over thirty thousand miles of track, two-thirds of which lay up north, on Union soil. This gave the Union an important edge, because its generals were able to transport and deploy troops much more quickly than their Confederate opponents. As a consequence, the American Civil War was the first war in which railroad infrastructure played a crucial part. It had taken all of four decades for railroads to go from a curious attraction to a mainstay of military planning and the economy as a whole.

By the time France fell prey to the turmoil of its great Revolution in 1789, a less flamboyant but even more momentous storm had been gathering over the West for more than half a century. This was the Industrial Revolution, which had begun to turn a world dominated by small-scale, workshop-based, demand-driven production into one of economies based on large-scale industrial production in specialized mills and factories. As a result, the mainly rural societies of Europe, where over 90 percent of people were employed in the agricultural sector and provided for their own needs, grew steadily more urban in character. Production became supply-based, and trade, rather than self-sufficiency, was the name of the game. In this new world, most people would spend their days working on a single product, a commodity they themselves rarely needed or could even afford to buy.

The Industrial Revolution first took off in the English midlands halfway through the eighteenth century. Until then, spinning and weaving had been typical cottage industries. All that began to change soon after the invention of the flying shuttle by a weaver called John Kay in 1733. This increased the speed of weaving so much that, to keep a single loom busy, more and more hands were needed for spinning and feeding in the yarn. This in turn led to the mechanization of the spinning wheel, culminating in 1780 with the invention of Crompton's mule, a spinning machine capable of producing both delicate silk yarn and more robust thread at high speeds. Meanwhile, mechanical power, based first on water and then on steam, was gradually taking the place of old-fashioned muscle. Ever-larger numbers of spinning machines and looms were concentrated in great halls around a single, central steam engine. By 1785, more than one hundred thousand mechanical looms were clattering away across the British Midlands. The factory had been born and, with it, mass production.

Weaving provided early capitalist industry with its organizational standard, serving as a model for the modernization of more and more other trades. By 1850, the number of people employed as factory workers in industrial towns had already surpassed the number of those who made a living in the country. Once more, a few decades were all it took to go from the initial invention—in this case, the flying shuttle—to a completely new way of manufacturing goods, with all the new tools and skills and even organizational principles that this entailed. And this model quickly came to dominate production in all but the most isolated, sparsely populated regions of the world. In fact, most of today's industries are still based on these same organizational principles.

* * *

The same pattern—a technological breakthrough followed by a rapid transformation from idea to useful, practical product, which then spreads all over the world—could be observed long before the Industrial Revolution got under way. Movable type, for instance, was invented in the West around 1450 by Johannes Gutenberg of Mainz. We now know this as the mechanical printing press, which used interchangeable stamps for each individual letter, all cast from a lead alloy.

This was an enormous improvement over the existing woodblock printing technique. In woodblock printing, the imprint of a complete page was carved out of a single block of wood, which was then used as a stamp. This procedure was extraordinarily laborious and difficult, requiring exceptional carving skill to arrive at a readable result. As a consequence, woodblocks were mainly used for detailed illustrations, leaving ordinary text in the hands of traditional human copyists.

Gutenberg was not the first to think of breaking up the block into individual characters. The Chinese had developed movable type some four centuries before, using porcelain letter stamps. But they needed enormous stocks of stamps, for there were thousands of characters in the Chinese writing system, and any character might occur any number of times on a single page. This made the procedure both very costly and hopelessly inefficient.

Gutenberg, however, had the good fortune of living in a world in which scripts were alphabetical, comprising no more than a few dozen different characters and punctuation marks. And he came up with another brilliant idea: rather than making all possible character stamps and keeping them indefinitely, he cast the stamps using a mold as and when he needed them, and then recycled the lead alloy. So in principle, he needed to keep no more than a few dozen molds and a bucket of alloy on hand at any one time. That way, printing was not only much faster than the old method of copying texts but also far cheaper.

Gutenberg's lead type opened up unprecedented opportunities to spread knowledge and ideas. From then on, producing a page of printable text was not only relatively quick, easy, and cheap but also yielded a much more consistent result. As usual, one improvement followed the other in quick succession. Gutenberg developed new, oil-based inks, allowing for the production of still more copies of good quality. Print runs grew to previously unimaginable sizes. And the number of works in print went through the roof: fliers, pamphlets, poems, political satire, diatribes, squibs, essays, songbooks, homilies, and works of fiction soon flooded countries with alphabetical scripts, inspiring ever more people to learn how to read. The jewel in the crown of the printing press, of course, was the gazette or newspaper. The spread of knowledge of current affairs to the farthest reaches of society would forever fundamentally change the balance of political power.

Despite the best efforts of the powers that be to control and curb the exchange of printed ideas and information, the printing press changed the Western world so deeply that it was hardly recognizable just a few centuries later. By the beginning of the twentieth century, the medieval fiefdoms, bishoprics, principalities, and kingdoms where rulers commanded loyalty by wielding absolute power on God's authority had all but disappeared, giving way to more or less modern constitutional monarchies and the odd republic. Today, another hundred years later, only two dozen monarchies survive worldwide, the vast majority of countries having become republics of one kind or another. Likewise, many traditional, class-based societies have evolved into more democratic, representational systems in which the separation of church and state is more or less taken for granted.

And, of course, the people living in these countries have changed as well. From a largely illiterate, ignorant, disenfranchised, and sickly mass of paupers, the subjects of old have developed into increasingly better organized, better informed, healthier, and more prosperous citizens. Naturally, there are still disadvantaged groups; the underclass, in a sense, lives on. But even they are much better off and enjoy far greater security than ever before. And these days, they are the exception, not the rule.

Thanks to the printing press, science and technology began to flourish like never before, changing our existence in more ways than we can count. But during the almost five hundred years of its existence, the mass-market printing and typesetting technology itself changed little. Like so many other production processes, it ran on steam from the eighteenth century, but for radical innovation it would have to wait until the late 1800s. Only then did linotype arrive on the scene, which in turn was replaced by offset lithography during the 1960s and, by the end of the century, digital, computer-based printing.

* * *

The further back we go, the more events seem to be shrouded in the mists of time. In the relatively empty and unorganized world leading up to the Middle Ages, natural, political, and cultural barriers were huge obstacles. It took more than a thousand years for the jealously guarded secret of making paper to reach Europe from China around 1200 CE, having first been leaked to the Arab world. Likewise, those who lived in the Americas, sub-Saharan Africa, and Australia remained almost entirely cut off from the developments of Europe and Asia until after adventurers like Vasco Da Gama, Christopher Columbus, Ferdinand Magellan, Abel Tasman, and James Cook made their great voyages of discovery. But providing that oceans, wars, and isolation did not get in the way, the times were always changing, and fast—right down to the invention of the wheel.

We will never know exactly who invented the wheel, but once it had first been used somewhere in the Middle East five thousand years ago, it popped up everywhere on the Eurasian continent in a matter of centuries. Knowledge of how to make wheels and put them to good use must have traveled through that sparsely populated, roadless world at an average speed of about fifteen miles per year. Under the circumstances, this was breakneck speed.

Meanwhile, the first writing systems were starting to blossom around Asia. Although writing was invented independently at different times in different places, there is no doubt that almost every alphabetical system ever used was derived from the very first writing system. This was invented by the Sumerians, who inhabited the fertile flats around the Tigris and the Euphrates, in what is now Syria and Iraq. Not surprisingly, it took a few thousand years for the first fully functional alphabetical script to develop out of the earliest attempts, which relied largely on pictographs. Writing was first used in the fledgling arts of arithmetic and bookkeeping: it was a handy tool for traders and accountants, and for people who wanted to assert ownership over things and keep track of their possessions. In time, however, people began to see how their signs and markings might be put to broader uses. How they might serve to record things that had nothing to do with arithmetic or bookkeeping at all, like poetry and prayers, curses, and declarations of love.

The cuneiform script that the Sumerians eventually came up with still showed signs of its provenance, with crude little pictures of concrete objects like a pot, a goat, or a bushel of grain. But by and by, the Sumerians began to think of their pictographs as representing words for objects, rather than the objects themselves. Instead of representing a particular thing, a pictograph now symbolized a sound pattern linked to a mental concept. This in turn triggered a process of simplification, stylization, and further abstraction, which opened up a plethora of new possibilities. A sign that once stood for a *pot* and then for the sounds *p-o-t* could eventually be used to indicate the same sound pattern as part of a word that had nothing to do with pots or pans, like des*pot.* At that point, the pictographic code of old had turned into an abstract, syllabic writing system.

A syllabic writing system has two major advantages over a purely pictographic system. First and most importantly, because it is an abstract system it can be used to "talk" about anything, not just the things you can see and point at. It also needs fewer signs—but on this score a lot more can be gained by linking signs not to *syllables*, but in an even more fine-grained way to individual *sounds*. Depending on the language, a syllabic system may need anywhere from fewer than a hundred to more than a thousand symbols, whereas alphabetic writing systems often get by with fewer than forty, punctuation marks and diacritics included.

The invention of writing required considerable powers of abstraction and a great number of small but significant steps. People had to

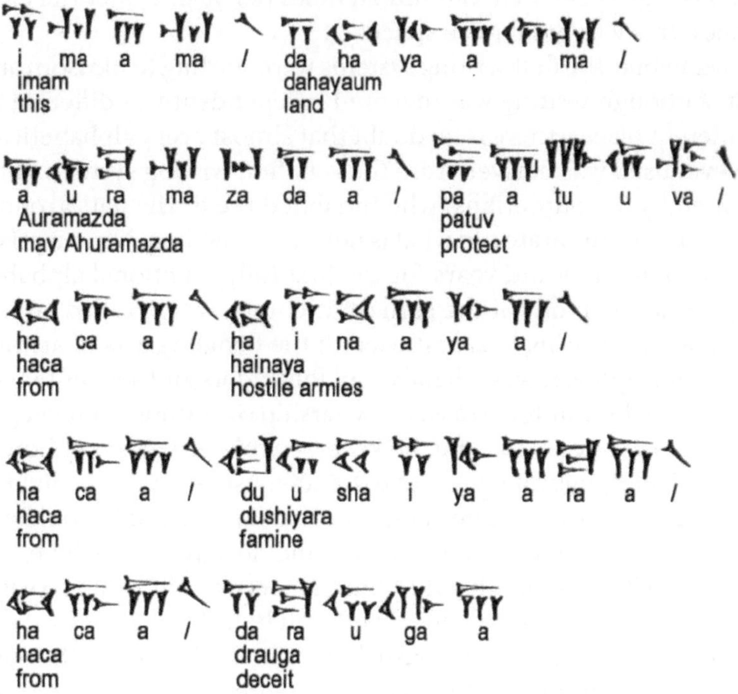

A cuneiform text: May Ahuramazda protect this land from the enemy, famine and deceit.

change their view of what symbols meant and how they related to things, sounds, and sound patterns. This cannot have been easy. Even today, many people find it difficult to understand and accept the abstract notions of basic mathematics. So it must have been a slow, more or less accidental process, progressing in fits and starts. And although writing spread around the world relatively swiftly after its invention, just like all those other great inventions, it did so in a slightly different way. The *idea* of writing gained ground fast, but not necessarily the cuneiform script itself. Many peoples developed a script of their own from scratch. Some came up with a more or less pictographic system, like the Chinese and the Egyptians, others with a syllabic or an alphabetic writing system, and still others with combinations of all three. The result was the wealth of different, mutually unintelligible writing systems still in existence today. At first glance, this seems strange and inefficient. But on closer inspection, it was only to be expected.

Once they become available, things like computers, cars, trains, steam engines, industrial production processes, printing presses, and wheels work equally well for everyone. They can be adopted "off the peg" and put to use immediately. Writing systems are not like that. They are bespoke. Even between closely related languages that use the same basic alphabet, there can be considerable differences. The nations of the Americas and Western Europe all use the same Latin-derived alphabet. English and Dutch make do with the common core, but German has the extra signs *ä, ö, ü,* and *ß*. French uses more than a dozen vowels marked with ´, ` and ^, in addition to *ç*. The Spanish invented their own peculiar *¿, ¡,* and *ñ*, and the Portuguese added *ã* and *õ*. In Scandinavia, there are also *å* and *ø*, and in Iceland *þ* and *ð*. The Turks, when they adopted the Latin alphabet in 1928, felt the need to add *ç, ğ, ö, ı,* and *ş*. Moreover, even English makes use of specific letter combinations to denote individual sounds, like *sh* and *th*. Dutch has its *ch, ij, au, ou, eu, ui,* and *oe*, and Italian invented the triplet *cch*, pronounced as k.

In short, there is no neutral, one-size-fits-all alphabet. Even the system of symbols designed by the International Phonetics Association (IPA), with its more than 140 different symbols, fails to cover everything. For everyday use, the IPA alphabet is utterly impractical. Languages usually have no more than about fifty sounds. Why learn a hundred symbols or more for sounds you will never use, some of which you cannot even produce? Moreover, alphabets are tailored to represent only those differences between sounds that distinguish one word from another. A hamburger is always spelled h-a-m-b-u-r-g-e-r, even though a weathered Texan pronounces the word very differently from a glib Boston lawyer or a Geordie from Durham, England. Writing systems have a considerable degree of abstraction and simplification built in, which actually enhances readability by stripping away regional and social differences in pronunciation.

What is more, writing systems differ from other human inventions psychologically and emotionally. Our language is so much a part of our identity that it would feel wrong to force it into a different alphabet just like that. This is another reason why every society that came into contact with the idea of writing immediately set out to create a writing system of its own, be it a variant of what they came into contact with or a completely new system.

Not every society fell for the attraction of writing. The Germanic tribes that lived across the river Rhine from the Roman Empire mostly did without it until well into the Common Era, as did the Scandinavians.

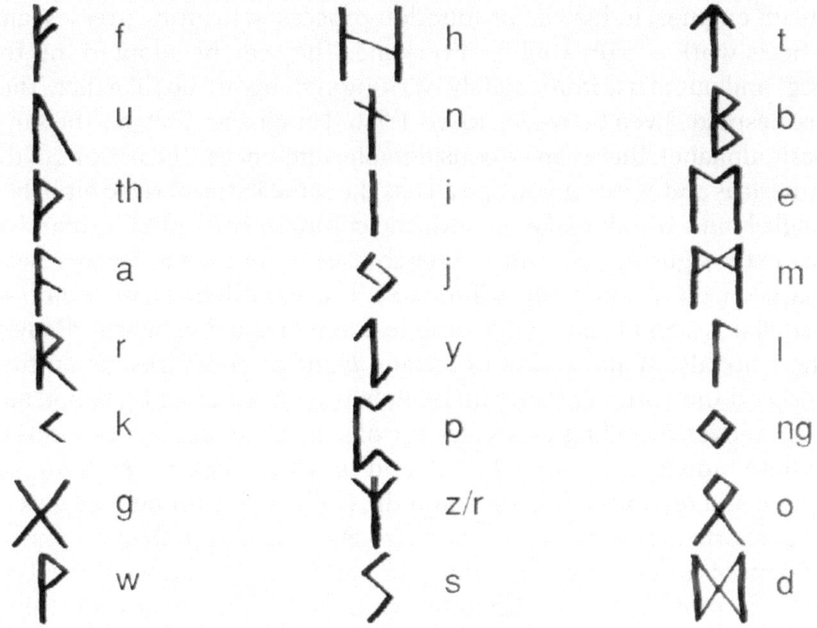

The runic alphabet, called Futhark after its first six letters.

They did have a semistandardized, runic alphabet, but all they used it for was to mark objects with their owners' names and make ritual inscriptions. The oldest examples of runic script date back to the second century CE. Other peoples forswore writing altogether, having no need for it. Writing holds little interest for a society until there are inventories to be made, laws to be encoded, or formal contracts to be drawn up. A culture also has to be sufficiently advanced to support specialized scribes or clerks. Many peoples did not bother with all this, and never would. Some such illiterate cultures still exist in the highlands of New Guinea, parts of Africa, and the Amazonian jungle, although chances are the last of them will succumb before long to the pressures of missionaries, miners, and loggers.

* * *

The oldest great invention the birth and spread of which we can more or less pinpoint is one that changed human life more than any other: agriculture. Just like the art of writing, agriculture emerged independently in different places at different times. But once again, it was the inhabitants of Mesopotamia, on the fertile mudflats around Tigris and Euphrates, who were the first. More than twelve thousand years ago, the climate, soil, and a rich range of flora conspired to create ideal circumstances for agriculture to arise in what is aptly known as the Fertile Crescent.

Once invented, agriculture spread quickly. Within fifteen hundred years, people were farming on the banks of the Indus River. After another fifteen centuries, there were people tilling land on the west coast of France and permanent settlements in eastern Spain. This peasant lifestyle, in other words, spread both east and west of the Fertile Crescent at an average speed of two-thirds of a mile per year. This may not sound very impressive, but what it boils down to is that the overwhelming majority of hunter-gatherer societies that came into contact with the idea of farming adopted it within a single generation. Just do the math: twelve thousand years ago, the world population numbered roughly ten million, with an estimated six million in Eurasia. Western Eurasia, the area between Calcutta and the shores of the Atlantic, was probably home to about four million people, all hunter-gatherers, who lived in bands of around a hundred individuals. So there would have been some forty thousand of these groups. They would have been concentrated mainly in relatively warm and wet regions, most likely on a strip of land measuring about 1,250 miles wide and stretching 6,000 miles from east to west. This means that the average band held a territory of roughly 300 square miles. Taking these territories to be more or less circular in shape, each would have measured about eleven miles across. At an average speed of two-thirds of a mile per year, a new band would be reached in every direction every eighteen years—roughly the amount of time it would have taken for an infant to grow up and produce its own first child.

So the speed with which people were willing to upset their traditional way of life in favor of farming is truly amazing, and even more so considering that not all groups would have been willing or able agriculturalists. The tribes of the Baltics, for instance, clung to their old lifestyle for more than a thousand years after their neighbors to the southeast had become farmers. A fair number of attempts must have ended in disaster, and hostility between groups as well as geographical and other impediments would have hindered contact and the transfer of ideas. But in spite of all these difficulties, this first great invention generally went the way of all the great inventions that would follow

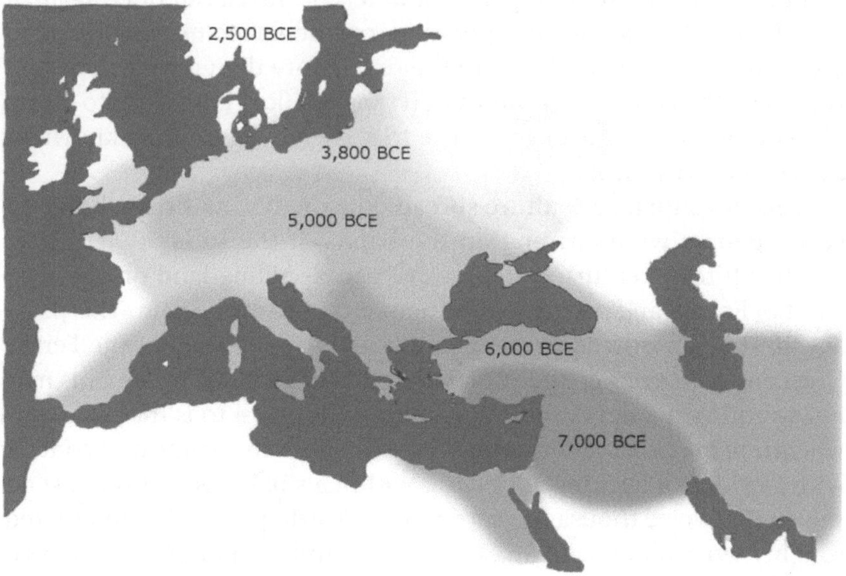

The spread of agriculture from the Fertile Crescent into Europe between 7000 and 2500 BCE. (after Diamond 1997)

it, spreading rapidly and being eagerly adopted far and wide, until it reached the most far-flung corners of the earth.

* * *

There was no reporter there to tell the world about it. No camera that could record it. And yet, the invention of agriculture was the greatest, most fundamental revolution of them all. Before that, humans were at the mercy of fortune just as much as any other creature that walked the earth. Climate, competitors, and germs kneaded and shaped the human species as relentlessly and capriciously as the wind grinds away at rock, ultimately polishing it down to dust. Agriculture changed all that. From the very moment it was discovered, it was man who, tentatively at first, began to shape nature.

Now, after twelve thousand years of tilling the fields and breeding stock, we have become a formidable threat to the flora and fauna around us. It did not look that way in the beginning. All agriculture did was make people a little less dependent on the forces of nature and the seasons. But there was a price to be paid. Farm animals brought with them new diseases, and people came into closer contact with pests and vermin.

Life on the farm cannot have been much easier than the old hunter-gatherer existence, but on average it must have offered better chances of survival. More importantly, the new, sedentary lifestyle paved the way for more and more varied ways of making life worth living. It was the gateway to almost everything we think of today as falling under the umbrella of culture and civilization. And, lest we forget, to the endless variety of everyday things and conveniences that are now commonplace, but which were completely alien to our hunter-gatherer forebears.

The big question is whether this momentous development would have been possible in the absence of a full-fledged language faculty. Could agriculture have come about without language, the way the Cro-Magnon cave paintings did? Would imitation and general intelligence have sufficed? To answer this question, we need to take a closer look at exactly what lay behind the "discovery" of agriculture.

Twelve thousand years ago, the climate in the Fertile Crescent was becoming warmer and wetter, creating an ideal environment for wild cereals like barley and Einkorn wheat. Vast fields of these cereals must have sprung up all over the region, so rich and so easy to pick that the locals could hardly avoid ending up with a surplus of grain. That was a problem, because their nomadic way of life did not allow them to lug all that grain around. But they were not keen on leaving it unattended either. So they did what Muhammad would do thousands of years later, when the mountain refused to come to him. If the grain could not come with them, they would stick with their grain, camping by their supplies for as long as possible. As they got better at storing their provisions and protecting them from wind, water, and vermin, they would have been able to stay in one place longer and longer. And they now had to, because after all, the storage pits they built needed protection. These may have been little more than crudely covered holes in the ground, but they were also the very first real estate.

Thus the knife cut both ways. On the one hand, their plentiful grain stores removed the need to move about, constantly seeking out new fields to pick. On the other, it robbed them of the freedom to do so. Before too long, their nomadic shelters must have given way to sturdier, semipermanent huts. In turn, this meant their tools no longer needed to be light and portable. Even a semipermanent abode would have been suitable for storing the heavy mortars and grindstones they used to turn grain into flour, which was much easier to digest.

As long as wild grains continued to grow in the vicinity of these primitive settlements, their inhabitants wanted for nothing. Season after season, the grain they collected was their staple diet, supplemented with meat and fish. But this was not real agriculture yet, not by a long stretch. All they did was take from the plenty that nature happened to provide. As soon as the circumstances deteriorated even slightly, or the minute more mouths were born than they were able to stuff with local grain, the game was up. Then the tribe had no choice but to return to its nomadic way of life, going off in search of new pastures.

Genuine agriculture only comes about when people start actively influencing the produce of nature by purposefully sowing and improving their crops. They can do this in two ways. Manuring, irrigating, and turning the soil will improve the growing conditions for the seeds. And the seeds themselves can be improved through selection and cross-breeding.

Selection basically means saving the grains from the best plants for sowing when the next season begins, and eating the rest. Over time, the seeds will develop into a mixture in which the best performing, richest and most resilient strains are vastly overrepresented. A much more intrusive way of meddling with nature is selecting certain properties by means of cross-breeding. This technique allows farmers to combine the properties of different strains of a plant into a single, new variety, or to eliminate less desirable peculiarities. This is how people create plants more suitable to their purposes than nature itself provides. In their very makeup, cross-bred varieties are essentially different from their wild ancestors.

* * *

But no agricultural technique makes sense in the absence of purposeful sowing. Preparing a good, rich soil is hard work. Nobody in their right mind will go to that much trouble unless they have a clear idea of the particular crop they want to grow on it. On the other hand, only a fool would leave the fate of his precious seeds to chance. A smart peasant sows his best seeds in well-prepared soil.

But how did people hit on the idea of actually sowing seeds in the first place? To come up with an idea like that, you have to grasp the relationship between the grains in a ripened ear and the fledgling plants that sprout months later. This is not at all obvious, least of all to seminomadic hunter-gatherers who make for new grounds whenever the wash gets poor. At best, they might have noticed upon their return the next season that new grain had grown in the same places. But this is so unspectacular an observation that it cannot exactly have been a light-bulb moment. Grain just happens to grow in specific places, as do berries and mushrooms and lots of other things. Had anyone actually stopped to think about this, they would have concluded that plants simply sprout from the soil itself, like flies were long believed to originate spontaneously in rotting meat. Or they might have attributed it to the hand of some mysterious, divine creator.

Some scholars maintain that the people of the Fertile Crescent learned about the role of seeds from their dung heaps. Hunter-gatherers often follow a more or less fixed itinerary, going from one semipermanent campsite to the next as the seasons demand. Each site would have a dedicated space for discarding leftover grains and other scraps, which was also the place where they would defecate. At some point, the group would break camp and move to the next site, and upon their return, lo and behold, there would be an exceptional abundance of wild grain on the old dung heap. This, so the story goes, would have opened their eyes to both the role played by seeds and the advantages of using fertilizer.

But would it? It is true that dung heaps make excellent breeding grounds for all sorts of plants, but there cannot have been much edible grain in there. Gathering wild grain is tedious, time-consuming work. So it is highly unlikely that people would have casually thrown away serious amounts of grain, especially of the potent uncooked variety. Even in the best of times, food must have been a relatively scarce commodity, not to be treated cavalierly. Then again, if grain happened to be abundant for a change, chances are that the gatherers either stopped working as soon as they had picked enough to cover their immediate needs or carefully stored the surplus in storage pits or huts. Throwing perfectly edible food onto a dung heap is surely the sort of wasteful behavior that only pampered, Western city dwellers like ourselves could come up with.

Even if the hunter-gatherers did squander precious food, how would that have helped them discover the relationship between the filth of April and the feast of August? Dung heaps had existed as long as hunter-gatherers had been around to produce them. So if all you had to do to hit on the idea of a seed developing into a plant, was to observe a dung heap, how could our ancestors have missed it for more than thirty thousand years?

The dung heap theory simply cannot explain the discovery of agriculture. Moreover, its proponents fail to take into account the most significant change in people's way of life in the Fertile Crescent. When the exceptionally generous climate of the day allowed them to abandon their nomadic lifestyle in favor of a more sedentary existence, they began to hoard.

Stores of grain must be kept under lock and key, safe from the elements and out of reach of thieving animals like mice, rats, and birds. Even today, keeping seeds, grains, and tubers in good condition for any length of time is no easy task. So back then it must have been even harder. All they had were crude storage pits, which were little more than rough holes in the ground. There was next to no stone pottery to speak of. The first earthenware, suitable for making pots that could actually be sealed, did not turn up until eight thousand years ago in Çatal Huyuk in Anatolia, the oldest real town ever discovered.

Down in the storage pits, apart from being eaten or damaged by vermin, two things can happen to stored grain. It may rot, or it may sprout in an untimely manner. It does not take much for either of these to come about. Just a little too much moisture and a touch of warmth will do the trick. So it cannot have been particularly unusual for people to return to their stores only to find a disappointing tangle of moldy runners instead of the lovely golden grains they had put in, or, alternatively, healthy seedlings that had sprouted from the stored grains. And it is this that may have spawned the idea that plants in fact sprout from grains—that if you want new grain plants, all you have to do is scatter some grain around. If this theory is correct, it was storage problems that led to the discovery of the relationship between grain and plant, and from there to real agriculture, that is, sowing for the express purpose of harvesting.

* * *

This storage explanation seems plausible, and it has the beauty of simplicity. Yet it cannot be the whole story. The fact is that discoveries of this kind can only come "automatically" to creatures of fairly advanced intelligence, in other words, to people like us. You can make any animal you choose stare at a bunch of seedlings for as long as you like, and it will never hit upon the relationship between seed and plant. It will never conclude that it could come to possess a lavish heap of food just by sowing and waiting. Even species with ample experience at stashing away nuts and seeds for later, like squirrels, have never taken a single step in the right direction. It is well over the head of even the smartest bonobo. So our ancestors of twelve thousand years ago must have shared with us certain mental capacities that were beyond all other species. And these powers enabled them to develop the lines of reasoning that would lead to the discovery of agriculture.

So what were these powers? One of them must have been a sense of individuality. To understand how a seed placed in a pot—or a storage pit—is related to the seedling found there later, you have to understand that they are one and the same thing. Strange as it may sound, you have to see the grains as individuals, each with its own identity.

This requires a well-developed sense of the flow of time, not just an awareness of duration, of "long" and "short," but a sense of time as a separate dimension. It also requires an understanding of the related notion of permanence: the ability to understand that the things you can see around you have a past and a future. A history, so to speak, during which an object may change in appearance or behavior, but nonetheless remains essentially one and the same thing. Then, and only then, is it possible to hit upon the idea that the seedling you find in a storage pit is the *same thing* as the grain you put in there weeks or months ago, and that the particular grain you put in that particular pit stayed there all that time, even when you were not looking.

To get a handle on all this requires a certain command of logic. You have to be able to comprehend the link between causes and effects. This, too, is closely related to a grasp of time and notions like "before" and "after," because effects follow their causes, not the other way around. An effect cannot arise if something else—its cause—has not taken place at some earlier point in time.

The painted cave walls of the Cro-Magnons bear witness to the hesitant burgeoning of our sense of individuality and identity. It is there, but only vaguely, a potentiality not yet fully realized. This sense of individuality is also a cornerstone of the human language faculty. We see it in the irrepressible urge that develops in every young child to label things, which is how we acquire our lexicon. Everything must be given a name, and thanks to these names, things acquire an identity by which they can be distinguished from everything else. Naming is the mental principle we use to order and manage our world.

Identity and individuality are not the same thing. Children do not stop at labeling the individual objects they come across; instead, they learn that names may also apply to *groups* of things with shared characteristics. From the moment they become aware that the quadruped of the household is called a *dog*, the concept *dog* is established in their psyche. And in no way is it limited to the individual animal with which they are familiar. They immediately start generalizing the idea, calling everything remotely resembling a dog *dog*. Horses, cats, cows—all these have to wait until children start differentiating between them and dogs in order to be graced with category labels of their own. It does not matter *what* label is used. If children are told that the dog is called *Fido* or *woof*, they will start calling all dogs and vaguely similar quadrupeds *fidos* or *woofs*. The point is that they immediately give all labels the broadest possible interpretation. We classify and categorize the world first in broad brushstrokes, and then embark on a lifelong journey of refining, subcategorizing, and reclassifying.

The overwhelming majority of words are merely sonorous representations of concepts residing in a person's mind (or visual representations, in the case of written text and sign language). These concepts are part of our inner world. They exist independently of our physical reality, although that reality was instrumental in forming them (or in the case of inborn concepts, at least in naming them). It is this independent status of our inner world products which allows us to grasp the connection between a seed and a seedling without keeping a constant watch over its development, because although the grain itself dramatically changes in appearance, its mental representation in our inner world remains constant. So it follows from the discrepancy between the seeds stored and the seedling retrieved that the concept "grain" manifests itself in the world outside in two very different guises, which is tantamount to understanding that seeds become seedlings.

The kind of abstract, logical reasoning required to do this is typical of language-driven thinking. One of the prime characteristics of human language and, as far as we know, human thinking, is that we think and converse about things that are not actually physically present at the time. They may be elsewhere, out of sight. They may be things of the past or the future, as in the case of the seeds and seedlings. Or they may not exist in the real world at all. No ape that was taught language ever showed the slightest sign of doing the same. So we must conclude that our powers of abstraction, just like our sense of individual identity, are a feature of our language faculty, or at least inextricably bound to it. Without language, abstract thought simply cannot exist.

* * *

The invention of agriculture was the very first cultural achievement that would have taken more than a good memory, some general intelligence, and a knack for imitation. Though it was discovered independently at several different times in several different places, the primordial peasants of the Fertile Crescent were the front-runners. This made them the first creatures on earth to prove that they possessed a full-fledged, completely operational language faculty. In other words, they not only looked like us but could also think and speak like us, and were mentally ready to invent all the trappings of culture and civilization we know today.

This tallies with another revolutionary innovation that occurred in roughly the same region. Around fourteen thousand years ago, the Natufians of the Levant began producing figurines with faces. Apart from the few much older, mysterious settlements in the Czech Republic and on the Russian plains, this was the first time in history that complete images of humans appeared. These same Natufians were also the first to produce explicitly erotic imagery, starting a tradition that is still with us to this day. These innovations coincided with the appearance of the first semipermanent settlements in the area and the start of another time-honored cultural tradition: ritualized burial, including votive offerings.

The first images of humans on rock faces in the southern Ural, at Ignatievka and Kostenki, date from the same period. At Ignatievka, images were found in 1980 of a man and a woman with 28 red dots between her legs. At Kostenki, depictions of females were also adorned with red smears. These are thought to be references to the menstrual cycle. If so, this indicates that their makers could not only count but also that they were aware of duration, regularity and cycles, and past, present, and future.

Sculpting and face painting, erotic art, solemn burial rituals, and signs of an understanding of relatively long cycles all point to the blossoming of a sense of individual identity, the emergence of a real "self." They indicate that a truly modern mind had emerged, one with a fully operational language faculty, which in turn opened the gateway to the discovery of agriculture and everything that followed.

The rest, as they say, is history. Relatively shortly after this new, language-driven mind had sparked mankind's first great discovery, agriculture, all the major farm animals were domesticated. Settlements appeared, initially semipermanent, and later developing into villages and towns with full-blown social stratification and functioning governments. With the sole exception of Sunghir in Russia, none of this had ever been seen before. The wheel followed, then writing, money, and an ever-increasing stream of technical inventions and cultural innovations. And this stream is still gaining strength and speed even today.

The question remains why agriculture was not invented much earlier. At times during the two hundred thousand years since the offspring of Black Eve began colonizing the earth, there must have been other places where the climate was as benign and wild crops as rich as in the Fertile Crescent twelve thousand years ago. Yet during all that time, nothing happened. And why did the Cro-Magnons, sophisticated cave dwellers that they were, never hit upon the idea of making storage pits? The only possible explanation is that the necessary mental machinery was not yet available. The truly modern mind, together with its language faculty, developed not long before all these things were invented. It took eons for all the building blocks of that mind to emerge, each for its own evolutionary reasons, and then another long stretch of time before all those components came together to form an integrated mind, unique in its power. But once in place, it went the way of all the great inventions it would make possible. This new, individualistic mind immediately put itself to work at full speed, and civilization began.

Epilogue

Echoes of the Past

To a person, a hundred years is a whole lot of time. A thousand years is a veritable eternity. But in the history of life on earth, even ten thousand years passes in the blink of an eye. From this perspective, it was only yesterday that our species reached its present stage of development. So recently, in fact, that we may still be able to find traces in our behavior and propensities of times prelingual, echoes of a mental state that was once dominant but has now been buried deep down in the furthest recesses of our minds.

One such remnant might be the ease with which we "forget ourselves", as it is called. For many of us, it does not take much to lose control, if only for a moment, to let our emotions take over and lose the capacity to restrain and direct our actions. We are all familiar with that involuntary curse that slips out when we stub our toe or hit our thumb with a hammer. Just a little further down the road of rage is giving the finger to an inept fellow motorist or, worse, making an angry maneuver just to *show* them. We know it is dangerous and counterproductive, but sometimes we just cannot help ourselves. Think of the loving father in a traffic jam on a stifling summer day who suddenly lashes out in desperation at his whining children in the back seat. Or—not rage-related but comparable nonetheless—think of the pubescent girls who dissolve into helpless giggling at the slightest sign of male attention. These are all cases of the animal in us breaking out and temporarily taking over, using old animal forms of communication: growling, giggling, and posturing.

Another trace of prelinguistic behavior may lie in the pleasure we get from a sense of group belonging. We all know the dangers lurking there. A class full of nice, well-mannered kids can morph into a predatory mob that turns upon a teacher at the drop of a hat. As a group, they relish the thrill of the attack, although afterward, individual members often find it difficult to explain what was so intoxicating. On their own, people rarely vandalize a bus shelter or attack a random passerby. These are typical group activities, like rioting. A one-man riot is a contradiction in terms.

There are positive examples of group behaviors as well. Singing, for instance, is especially enjoyable when we do it together. Dancing, too, is something we prefer to do in the presence of other dancers. So is drinking alcohol. Downing a glass or two in a bar with friends is perfectly acceptable, whereas solitary drinking is seen as just a hair's breadth away from a drinking problem. Or think of watching a soccer match, if only on television. Many people prefer piling into a pub to watch the game on a big screen to the comforts of their own couch. If they do stay at home, at the very least they will have the neighbors over to enjoy the match in one another's company.

Losing one's identity is a major component of this kind of exhilaration. Blending in with a group comes with a kind of anonymity and loss of individuality that we experience as both liberating and sheltering. This goes for rioters, protesters, and loitering youths, but also for sports fans in a stadium who spontaneously join in a massive wave like a swarm of birds, and vacationers who sun themselves en masse on the beach like so many penguins brooding. Sometimes we even feel the urge to emphasize this loss of individual identity. It is no coincidence that carnival revelers dress up in such striking headgear—a mask obscures its wearer's identity and, in doing so, neutralizes distinctions of class or social position.

Our cavalier attitude toward our individual identity is most evident in our penchant for uniforms. A uniform brands us as one of the group, and at the same time obscures our individual personality. We become anonymous: we are no longer a person, but instead have a purely instrumental identity. In a uniform, we are transformed from a some*body* into a some*thing*, a functionary, identified by stars, stripes, and ribbons. This makes sense when we think about the military, for example. The concept of the modern army relies on interchangeability. Even among highly trained special forces, the unit always comes first. Under fire, soldiers will take risks they would not dream of taking in ordinary civilian life. It is not altruism that drives them, nor love of their fellow fighters, most of whom they would not know from a bar of soap. They do it for the collective of which they feel they are part. When push comes to shove, their individual survival is secondary to the survival of their unit. This is why the company's honor, symbolized by standards, berets, and insignia, is so important. Any veterans' parade is testimony to the sheer strength of this sense of belonging.

* * *

So in spite of all the sophisticated modernisms in which we take pride, our ways are not as different from those of our prelingual ancestors as we might think. In fact, it would seem that even the collective memory of those long-gone days has not faded away entirely—a memory of the world before human beings mentally stepped out of their natural backdrop and were gripped by a new sense of self-awareness. Residual traces of this memory seem to ring through in the *Iliad* and the *Odyssey*, the oldest known works of Western literature created by Homer around 800 BCE.

Create might not be the appropriate term in the case of the legendary blind bard. In all probability, he mainly recorded and combined existing stories, many of which were already very old, having been passed on by word of mouth for generations. Even Homer himself might be a myth. Quite possibly the name denotes just one of a collection of long-forgotten poets, rhapsodists, and writers we can no longer make out through the mists of time that shroud the Greece of some three thousand years ago. Even by the golden age of classical Greece, only a few centuries after he is supposed to have lived, Homer's identity was already a mystery. What is clear is that the two great epics ascribed to him not only shed light on Bronze Age culture but also lift a corner of the curtain to earlier times—much earlier—all the way back, perhaps, to the days when truly modern people first walked the earth.

At first glance, the *Iliad* and the *Odyssey* seem like two peas in a pod, as tightly bound to each other as the seasons of a television series. Yet experts have long been aware of the enormous differences between them. The story of the *Iliad* unfolds against the backdrop of a conflict between King Menelaus of Sparta and the powerful city of Troy (or *Ilium*, hence the title). Troy was an important trade hub not far from the Dardanelles in the northwest of what is now Turkey. The conflict had arisen over Menelaus's wife, the ravishingly beautiful Helen. Shortly after their marriage, Helen had eloped with Paris, a visiting Trojan prince. When the city of Troy refused to return Helen to her lawful husband, Menelaus formed a great coalition of Greek nations. The same intoxicating beauty that had tempted Paris into carrying off Helen enabled Menelaus to unite the quarrelsome and jealous Greek princes. Before her marriage, they had coveted her down to a man, and to avoid going to war over her, they had taken a pledge to respect the lucky suitor who would eventually win her hand. When the Trojans stole their collective idol, they kept their word: together they would bring Paris and his cronies to their knees. Armies were raised and a great fleet was formed, and before long the combined Greek forces landed on the shores of Troy. There they met with tough resistance. So tenacious were the Trojans that it took the Greeks ten years—and a cunning plot involving the infamous Trojan Horse—to finally vanquish the city and retrieve Menelaus's wife.

The *Iliad* covers little of what happened during those years of strife. It is the *Odyssey* that recounts the main events of the war and the story of the Trojan Horse. In the *Iliad*, the siege of Troy is merely the background against which the real story unfolds. At its core is the conflict between two equally headstrong and ambitious Greeks, the Mycenaean king Agamemnon and Prince Achilles of the Myrmidons. The siege of Troy has been dragging on for nine years when their rivalry erupts into a vicious tug of war, with devastating consequences, over an enslaved Trojan girl called Briseis. The *Iliad* is a tale of vengeance, of rage and mistrust, greed and pride, that is to say, hard-to-control urges and emotions, and their deadly effects.

The very first word of the very first line of the *Iliad*, *μῆνισ*, reveals the central theme of the story: "wrath." The same trick is used in the *Odyssey*, which starts off with the word ἄνδρα, a form of ἀνήρ: "man." This is a story about the trials and tribulations of Ulysses, not a jealous, testosterone-driven killing machine like Achilles, but a wily mental powerhouse. Ulysses is no larger-than-life demigod, but a self-sufficient, independent individual. He is a self-interested smooth talker, a shameless whitewasher, a merciless looter and thief, and a man who has few qualms about sacrificing even his most loyal companions to save himself. But he is also a diminutive creature, condemned to survive by his wits in a dangerous world. Cunning and subterfuge are his most powerful weapons. His curiosity and inquisitiveness make him strong and unpredictable, but also time and again lure him to within a hair's breadth of his own undoing.

* * *

In the *Iliad*, there is a fine line between the world of mere mortals and the divine world of the gods. People and deities mingle with one another freely. The start of the *Iliad* sees a frustrated Achilles pleading with his divine mother, Thetis, to intervene on his behalf with Zeus, whose support he needs against Agamemnon. Thetis promises to oblige, but not just yet, for the supreme god happens to be away on a business trip, attending a dinner party with the Ethiopians. On his return in a week or two, she reassures Achilles, she will do what she can to further her son's cause. There is no hint here of divine powers like omnipotence, omniscience, or omnipresence. On the front door of Zeus's ethereal palace atop Mount Olympus hangs a mundane sign of the sort "be right back."

Equally devoid of any sense of the divine is the behavior of Zeus and his wife, Hera, when he gets back from the tropics. Zeus gives in to Thetis's pleading but only grudgingly, to avoid getting in trouble with the missus. All right, he says, but please, would she make off now before Hera spots her? But his efforts are in vain, as jealous Hera keeps close tabs on her philanderer of a husband. She confronts him, and Zeus, sick and tired of being spied on, hits the roof and threatens to give her a damn good beating. In the world of the *Iliad*, even the divine first household is no stranger to domestic trouble.

This is by no means exceptional. Throughout the *Iliad*, gods, mythical beings, and mortals are in constant contact, both in spirit and, so to speak, in the flesh. The Olympic gods enthusiastically partake in the battles waged between the Greek and Trojan armies. Even the cuddly Aphrodite puts in an appearance, the goddess of love coming away bleeding from a wound inflicted by the spear of the mere mortal Diomede.

This is in striking contrast to the *Odyssey*, which follows the trials and tribulations of Ulysses after the fall of Troy. The journey back to his beloved island of Ithaka takes him a full ten years, and although the Olympians pop up frequently during this time, it is largely to raise storms and mountains, petrify ships, and play all kinds of other supernatural tricks. They rarely make direct contact with the mortals, and when they do, they come disguised as a friend, a stranger, or an animal. Some of them are out for Ulysses's blood, and others are with him, and while they happily offer their protégé advice, they largely refrain from interfering with the natural course of events. There are only two exceptions to this general rule. At one point, the messenger god Hermes hands Ulysses a bouquet of fresh herbs, which will make him immune to the magical potions of the witch Circe. And upon his return to Ithaka, Ulysses's great protector, Pallas Athena, momentarily reveals herself to him.

Ulysses, for his part, shows far less regard for the gods than would have been imaginable in the *Iliad.* He is not averse to brushing aside the gods' counsel aside to set a course of his own—although he pays the price for his cockiness. One day he has to sail his ship through a narrow strait right under the noses of the many-headed monster Scylla. The divine witch Circe entreats Ulysses not to challenge Scylla, but merely to slip by as fast as he can. It is sensible advice, but too much to ask of our irrepressibly curious, inveterate hero, who wastes precious time donning his shining armor and provocatively taking up a position on the bow of the ship—for "what else can a real man do?" Half a dozen sailors are snapped up into the waiting jaws of the monster, paying for his recklessness with their lives.

Individuality and identity are prominent themes in the *Odyssey,* often in the form of temptations leading either to a loss of identity or even death. When Ulysses and his men set foot in the land of the Lotus-eaters, they find themselves in a huge opium den *avant la lettre,* where all and sundry drift through life in a soothing haze of oblivion. It takes all Ulysses's powers of persuasion to get his men back on board before they start to lose themselves as well. In another scene, besides turning Ulysses's men into swine, the wicked witch Circe makes them forget who and what they really are. Under her spell, they fail even to recognize their commander. Eventually Circe agrees to lift the spell, but this is only the starting signal for a year-long feast of unlimited food and wine that might well have gone on forever had Ulysses not pulled himself together.

The *Odyssey* is full of tales like these. In yet another, the sweet singing of the Sirens drives passing sailors so mad with desire that they willingly wreck their ships on the rocks surrounding the Sirens' island. Circe tells Ulysses that earplugs made of wax will help his crew get safely past the honeytrap, and this time Ulysses wisely heeds her advice, having all his men plug their ears. But instead of plugging his own, he has his shipmates bind him to the mast, so that he can hear the irresistible chanting of the Sirens and yet live to tell the tale. It is a brilliant display of self-restraint, in the most literal sense of the word. Eventually, Ulysses somehow survives all his companions and winds up alone on the island of Calypso. Almost alone, that is, for there, in the bed of the alluring nymph, he almost forgets himself one last time. Only after several years does he manage to get a grip on himself and strike out on the last leg of his journey back to Ithaka.

Identity fraud is another major theme in the *Odyssey*. In the final, murderous scene, Ulysses kills off the whole band of forty spongers who have spent years trying to get at his faithful wife Penelope. The crux of this scene is that the returning lord of the manor is not recognized until it is too late—Pallas Athena has disguised him as an old beggar. Then, of course, there is the famous Trojan Horse Ulysses comes up with, yet another scheme of Ulysses and perhaps the most famous case of false identity in history. Left behind on the beach by the defeated Greeks, the wooden horse looks like nothing more than an innocent statue to placate the gods. The scheme works on the assumption that the jubilant Trojans will take it as a prize, towing it within the walls of Troy. Later, under the cover of darkness, the assault team concealed inside it will jump out and throw open the impregnable city gates to the Greeks, who had feigned sailing home in defeat. During the wait, Ulysses has a hard time keeping his men from jumping out prematurely and engaging in old-fashioned, open battle.

More than anything else, the Trojan Horse marks the juncture between the mythical, collectivist world of old and a more modern, individualistic, and essentially rational perspective on human existence. The old values are still there. The urge to do the irrational "honorable" thing rages within the horse's cramped confines. But logical Ulysses manages to control himself and his comrades until the moment to strike comes as planned.

Yet the apogee of Homer's masterpiece is not the horse but the episode in which Ulysses cannot contain his curiosity once again, and goes off to explore a cave with a dozen of his men. The cave turns out to be home to the Cyclops Polyphemus, a ferocious one-eyed giant, and his sheep. Upon returning with his flock, the giant blocks the narrow cave entrance with a boulder. Before long, he discovers the trespassers and leaves no doubt as to his intentions: he grabs two of the men, smashes their heads into a pulp against the walls of the cave, and gobbles them up for breakfast. Another two become dinner.

Faced with certain death, Ulysses comes up with a desperate ruse. When the giant takes his sheep out to pasture, he and his remaining men will escape by hiding under the bellies of his largest rams, holding on to the wool. But this will only work if Polyphemus does not spot them. So Ulysses proposes to put out the Cyclops's single eye with the pointed end of one of his own enormous shepherd's crooks. There is only one problem: what if Polyphemus makes so much noise that his neighbors come to see what is wrong? Before long, Ulysses is struck by a flash of genius. That night, as Polyphemus is getting ready to turn in, Ulysses strikes up a conversation with him. He asks the Cyclops for his name. "Polyphemus," comes the answer. "What's yours?" To which Ulysses famously replies, "Nobody. My name is Nobody."

When the pointed crook pierces his eye a few hours later, the giant starts hollering and shrieking as expected. Before long, the neighbors are gathering on the other side of the boulder. "Polyphemus, what's the matter? Is someone robbing you?" "It's Nobody," he yells back. "I'm being attacked by Nobody!" Mildly confused, but satisfied that nothing serious is up, the neighbors shrug and return home. "Perhaps his loneliness is getting the better of him," one says. "A bad dream," another suggests. "Midlife crisis," says a third. "Let's leave him be for a while."

At dawn, when Polyphemus is due to take his flock out, all the blind giant can do is feel the backs of his sheep to check there are no Greeks riding on them. His assailants, clinging to the sheep's woolly undersides, pass unnoticed. Ulysses's ploy was as ingenious as it was effective—but its real significance lies in the fact that it proves he was a truly modern man. To be able to forsake your own identity, you must both have your own identity and be aware that you have your own identity. You have to be able to view yourself as detached from the backdrop of the rest of the world. Moreover, in order to become "nobody," you have to be able to separate yourself from yourself, to be certain of your own continued existence even as you outwardly deny it.

And so Ulysses's little scheme marks the magnificent final step in the long, long journey the Cro-Magnons had kicked off so many years earlier, as they pressed their pigmented hands to a cave wall with a vague, amorphous sense of wonderment. The enormous brains of our ancestors enabled them to construct an inner world of unprecedented size and detail. But it was language that allowed them to structure that world and create all sorts of novel concepts and notions. These concepts help us think about the things that go on inside our own bodies and minds, instead of just passively experiencing them. Many of these concepts exist inside the mind alone, serving to order the abstract world of thought. "Economy," "democracy," and "justice" are good examples of these, but so too are notions like "fuel," "object," and "goal." Then there are logical concepts that are exclusively linguistic, some of which cannot exist in the physical world at all. This group is made up of those notions that merely negate or deny something, including "nothing," "nowhere," and, of course, Ulysses's infamous "nobody."

* * *

Was that really the end of it? Does the *Odyssey* truly mark the final step on the long journey from the primeval, animal darkness toward the enlightened condition of the modern human mind? Or is the mental dawn not a singular event but an ongoing process, one that will continuously, endlessly alter our position in the world? Is the way we understand the world and everything in it still changing? Although the emergence of the language faculty and the kind of self-awareness Ulysses showed were certainly momentous events in the evolution of the human mind, there is no reason to suppose it ended there. At first glance, it may seem to us that there have been no spectacular changes since. But it seems just as likely that this is merely a reflection of our complacency and our sloppy powers of observation.

If we look carefully, several fairly recent developments suggest that the human mind has continued to develop right up to the present day—and is still changing. These changes are easily overlooked, because they mostly affect what we instinctively know and understand and cannot help feeling. They are nonetheless real, and one of them has to do with gambling.

One of the world's oldest games of chance is knucklebones, which dates back to long before dice were invented. The bone used in the game was the *Astralagus*, a rock-hard, oblong-shaped bone from the ankle of a sheep or deer. It has two wide faces and two narrow ones, each of which had a fixed value, like the sides of a die. The game usually consisted of throwing a handful of knucklebones in the hope of arriving at certain combinations. Curiously, the Greeks did into take into account the obvious fact that a knucklebone is more likely to come to rest on one of its wide faces than on a narrow one. The traditional scores for the wide faces were 3 and 4, those for the more difficult narrow faces 6 and 1. Likewise, different combinations were worth different numbers of extra points, even though the chances of any of them being thrown were exactly the same. Although the classical Greeks played the game for many centuries, and certainly had some understanding of probability, none of them seem ever to have caught on to the design flaws in the game. Either they did not notice, or they failed to see their significance.

Throwing dice, which probably developed out of the old knucklebones, was equally ill-understood until only quite recently. Roman legionnaires happily gambled away all their possessions using dice so unevenly shaped that even a ten-year-old today would reject them out of hand. The same goes for playing cards. The popularity of card games went through the roof after the invention of the printing press in the fifteenth century, when they could be mass-produced cheaply. Before long, fortunes great and small were changing hands at the cards table. But however high the stakes might have been, for centuries nobody seemed to catch on to the fact that kings, queens, and jacks had a top and a bottom. From the way players turned their cards in order to read their hand, opponents could easily see how many "pictures" they held. It was not until the nineteenth century that symmetrical picture cards were invented.

Although few people these days are fooled by misshapen dice or asymmetrical playing cards, most of us have a grasp of statistics that is far from solid. It is still relatively easy to cheat people out of their money by means of a crooked game of chance. The most striking modern example is roulette. Even the most favorable variety of the game, the French version, is biased on account of its 0, which benefits the bank alone. Because of this, all players are destined to lose in the long run, whereas all banks will ultimately return a 2.7 percent profit. You do not have to be a statistician to understand this—yet we keep on handing over our money to smiling croupiers until *rien ne va plus*.

Statistics are great for pulling the wool over people's eyes. "There are three kinds of lies," Mark Twain had British prime minister Benjamin Disraeli say, "lies, damned lies and statistics." The statesman's exasperation exemplifies perfectly how helpless even the well educated in the nineteenth century felt when it came to notions like probability and risk. In many people's eyes, statistics was a modern kind of wizardry favored by those who were up to no good. Today, we seem hardly better off than Disraeli or Twain, but compared to classical Greece and Rome, something fundamental does appear to have changed. Slowly, very slowly, our minds seem to be acquiring an ever-better understanding of probability and risk.

Another example may be the advent of Expressionist painting in the late nineteenth century, followed by the rise of truly abstract art. By now, poor Vincent van Gogh has become the world's pet painter, something he could never have imagined. But the same cannot be said for Willem de Kooning, Anselm Kiefer, Karel Appel, or even Pablo Picasso—their works are still to some extent controversial. To those who reject such art, it is not even a matter of liking or disliking it. For them, it is simply not art. Yet there are others who derive just as much pleasure from abstract art as from any old master, from Cranach to J. M. W. Turner.

It is important to distinguish modern abstract art from the orderly, geometric patterns of decorative Arabian art, which never gives rise to the kind of controversy surrounding modern abstract work. Perhaps they remind us of the decorations humans have been painting on their pots and pans, and their own bodies, for more millennia than we can remember. One key thing they have going for them is their symmetry, a property that is very pleasing to the eye. By contrast, the kind of abstract art that arrived on the scene around the turn of the twentieth century made—and still makes—many people feel uneasy. They just do not know what to make of what they see as random chaos. Until roughly 1900, artists must have thought the same way; otherwise, at least some of them would surely have begun to produce abstract works at some earlier point in time. This makes abstract art a truly new phenomenon, the kind of fundamental innovation that may reflect a new phase in the development of the human mind. If we look beyond the snobbism and the received ideas that distort the picture of how people appreciate art, enjoying abstract art seems to be something not everybody is equally ready for. It may take a number of generations yet for this progressive mind-set to become standard.

A different global trend has been the slow but undeniable increase in the value of an individual's life. This can be seen in the changing attitudes to war casualties and workplace accidents, to name just a few examples. Only a century and a half ago, the enormous loss of life during the American Civil War was swallowed with relative ease. The battles of Shiloh, Gettysburg, and Chickamauga were but a few of the clashes in which upward of thirty thousand soldiers perished in a matter of hours. Over just four years, 750,000 men met with untimely deaths, a whopping 2.5 percent of the population. Proportionally, this is ten times the toll America paid in the two world wars, which at over half a million lives was still not to be sniffed at. Since then, such loss of life has become less and less acceptable. Thirty years after the end of Word War II, it took "only" fifty-eight thousand American dead in Vietnam to plunge the country into a mental depression that lasted for decades. Today, the United States counts its losses one by one.

The same is true for the rest of the Western world. The long-term trend is clearly one in which loss of life has come to be seen as increasingly unacceptable, not only in the military but in the workplace as well. Seafaring is a shining example of the evolution of standards. Barely a century ago, it was a fact of life that many of those who set out to sea would not make it back—not just humble fishing boats but also much larger cargo vessels and even men-of-war regularly went down, taking entire crews with them. And in the early modern age, if the galleons and schooners that plied the intercontinental waves did manage to stay afloat, fatalities caused by accidents, scurvy, or starvation were part and parcel of the profession. These days it is a different story. Even a lone fisherman in trouble in the South China Sea has an outside chance of making the front pages in Western Europe or the United States. Likewise, factories and industrial plants in the West are worlds apart from what they used to be. We have come a long, long way from what were considered normal working conditions during the nineteenth century, and slowly the rest of the world is following suit.

From a strictly rationalistic point of view, it is hard to see why we place so much more value on human lives than we used to. It is not as if human flesh is in short supply—in fact, there are far more of us than the earth is able to sustain. A strait-laced utilitarian philosopher like Jeremy Bentham would no doubt decide that society could spare more than a few of us. So something else must be going on. That something seems to be a new appreciation of the quality of life and the status of individuals. Barely two centuries ago, only a minority of people had any qualms about slavery, even in the richest, most powerful, and intellectually advanced nations. It would not exactly be pleasant to be enslaved yourself, of course, but few educated people considered slavery a moral problem. And only a couple of centuries before that, every imaginable degree of bondage and serfdom was seen as perfectly normal all over the world. Such views are now truly a thing of the past.

Roughly parallel to the emancipation of slaves went that of another group traditionally considered a commodity to be owned and used at will: women. We may frown upon the lamentable position of women in many parts of the world today, but we must not forget that only a few decades ago white European and American women were not much better off. And this state of affairs was considered the natural order of things by the great majority of men and women alike. In the West, most slaves had long been emancipated before the struggle for the emancipation of women took off in earnest.

First to grant women voting rights was the state of Wyoming, in 1869. Most other US states and most of the nations of Western Europe followed suit around the time of World War I. And over the course of the twentieth century, one by one, almost every other country in the world did the same. Women in Switzerland had to wait until 1971, and in one or two cantons they still do not enjoy the same voting rights as men. Women in many parts of the world have also gradually gained the right to open a bank account, inherit from their spouses, enter into legally binding contracts, divorce their husbands, and so on. And yet even in the West, only a generation or two ago, all this and much more was considered both daft and irresponsible.

Such revolutionary changes run deeper than mere politics. They reflect fundamental changes in the way we view ourselves, others and society in general—changes that seem to be irreversible, although they can of course be suppressed. Changes like these occur spontaneously; they are not instigated or controlled by any one individual or group. The suffragettes played their part in the early twentieth century, and a great many other activists followed in their footsteps, right up to the present day. And their efforts certainly had an effect: they got people thinking, changed people's views on the positions and roles of the sexes, and helped put an end to many discriminatory laws and practices. But they were as much the product as they were the promoters of the changes that were taking place. They gave voice and focus to what had been going on spontaneously in people's minds for generations. It was a slow, ongoing process of mental reorientation in man and woman alike, one that is perhaps best exemplified by the different outcomes of the two great wars of the twentieth century.

When the Great War broke out, millions of men were packed off to the trenches to fight, depleting the workforce. It was up to the women to take care of business at home. They staffed the abandoned stations in offices, workshops, and factories and, for four long years, produced and packed the bullets, grenades, and other tools of war used in the carnage at Ypres, Verdun, and elsewhere. But once the war was over and those who had been spared came home, the vast majority of these women gave up their jobs and went back to their domestic chores as if nothing had happened.

Only a generation later, another great war broke out in September 1939. For about eighteen months, the fighting remained limited to western and central Europe. Then Hitler invaded the Soviet Union and, six months later, Japan teamed up with the Germans and assaulted Pearl Harbor, forcing the isolationist United States to enter what had become a truly global conflict. Once again, men marched off to battle everywhere, and women duly poured into offices and factories to fill their shoes. But when this war finally came to an end, in 1945, women did not meekly return to their kitchen sinks. Instead, they claimed for themselves a permanent and independent presence in public life. They hung on to their jobs, and in the following years their younger sisters and daughters too came out in droves to seek employment. They did so of their own accord, as naturally as their mothers had given up their jobs a quarter of a century before. Now, some seventy years later, working women are the norm in large parts of the world and even high-ranking female officials, cabinet ministers, and presidents are no longer seen as somehow anomalous.

The emancipation of women is most visible in the Western world. The West is made up of a collection of rich, stable, democratic constitutional states with the highest educated levels found anywhere in the world. These states are by no means perfect, but broadly speaking they are places where civil rights and liberties are cherished and upheld. Such societies offer a lot of social and economic mobility and ample opportunities for spontaneous social experiments, much more so than countries paralyzed by tradition, poverty, corruption, and oppression. There, signs of changing patterns of thought are much less in evidence—but this is not to say those changes are absent. Consider Iran, for example. In 1979, a popular movement replaced the Shah's repression with an even more oppressive, totalitarian theocracy, catapulting the whole country straight back into the Middle Ages. Urban middle-class women, however, who were robbed of whatever freedom and personal dignity they had garnered under the Shah's more modern, Western-oriented regime, resisted this plunge into darkness as best they could. They dutifully wore their headscarves in public, but took care to let a wisp of hair escape from under them, a small but significant rebellion against the ayatollahs' efforts to impose absolute female modesty and subservience. Fifty years earlier, not one Iranian woman would have come up with a subversive idea like that.

Something similar has been going on in Turkey since the rise of President Recep Tayyip Erdoğan and his Islamist AKP, an acronym for Adalet ve Kalkınma Parti, Justice and Development Party. Large, mainly religious parts of the population are still willing to support him, but when he openly berated Turkish women for laughing immodestly in public in 2014, the female population hit back with extraordinary vehemence. Thousands went on record on Twitter, Facebook, YouTube, Instagram, and other low-threshold media, laughing out loud, squarely ridiculing their leader. A few went as far as posing topless with smileys painted on their breasts. It was a uniquely good-humored, witty protest, and extremely effective against the dour, mirthless precepts of the mustachioed president. It would also have been utterly unthinkable in Turkey even twenty-five years before.

Generally speaking, women around the world are becoming more confident and more ambitious. Even in countries like India, Pakistan, and Afghanistan, the oppression and disenfranchisement of women is slowly becoming an issue. And not just among women—there is at least a hint of awareness among men as well of the predicament of the other sex. Rape and violence against women no longer invariably go unpunished. While atavistic eruptions like the scourge that is Islamic State or the rise of sharia law in Indonesia may suggest there is neither progress nor hope for mankind, the long-term, global pattern is undeniable.

The emancipation of slaves and women may well be a natural continuation of the process that so many years ago turned the Cro-Magnon, still steeped in the collective, into a modern, language-powered individual, the kind of person epitomized by the wily, self-sufficient, and utterly self-centered Ulysses. But once the self had been discovered, the other gradually came into view. First the stranger, then the vanquished, and most recently the woman—slowly but surely, over many long centuries, the circle of those we count as "us" has been expanding. We could call this "progress." In reality, it is merely our current collective mind-set, the natural state of development of our inner world. It is simply what our minds happen to be capable of recognizing and acknowledging at this particular point in time—no more, no less.

Bibliography

Aitchison, Jean. 1989. *The Articulate Mammal: An Introduction to Psycholinguistics.* 3rd edition. London: Routledge.

———. 1997. *The Seeds of Speech: Language Origin and Evolution.* Cambridge: Cambridge University Press.

Arnold, Naomi, K.T. Ng, E.C. Jongman and P.H. Hemsworth. 2007. "Responses of Dairy Heifers to the Visual Cliff Formed by a Herringbone Milking Pit: Evidence of Fear of Heights in Cows (Bos taurus)." *Journal of Comparative Psychology* 121(4):440–446.

Barlow, Horace B. 1990. "The Mechanical Mind." *Annual Review of Neuroscience* 13:15–24.

Basalla, George. 1987. *The Evolution of Technology.* Cambridge: Cambridge University Press.

Becker, Ernest. 1971. *The Birth and Death of Meaning.* 2nd ed. New York: The Free Press.

Bernstein, Peter L. 1996. *Against the Gods: The Remarkable Story of Risk.* Hoboken, NJ: John Wiley & Sons.

Bickerton, Derek. 1990. *Language and Species.* Chicago: University of Chicago Press.

Bindon, Jim. 2006. "Distribution of Human Differences." www.as.ua.edu/ant/bindon/ant275/presentations/Distribution_of_human_differences1.pdf.

Binford, S. R. 1968. "A Structural Comparison of Disposal of the Dead in the Mousterian and Upper Palaeolithic." *Southwestern Journal of Anthropology* 24, 139–154

Bordaz, Jacques. 1970. *Tools of the Old and New Stone Age.* Garden City, NY: Published for the American Museum of Natural History by the Natural History Press.

Brooks Gruver, Rebecca. 1972. *An American History.* New York: Appleton-Century-Crofts.

Brown, Donald E. 1991. *Human Universals.* New York: McGraw-Hill.

Brown, Michael H. 1991. *The Search for Eve.* New York: Harper Perennial.

Burenhult, Göran, ed. 1993. *The First Humans: Human Origins and History to 10,000 B.C.,* New York: HarperOne.

Butterworth, Brian, R. Reeve, F Reynolds and Delyth Lloyd. 2008. "Numerical Thought With And Without Words: Evidence From Indigenous Australian Children." *Proceedings of the National Academy of Sciences* 105/35: 13179–13184.

Calvin, William H. 1983. *The Throwing Madonna: Essays on the Brain*. New York: McGraw-Hill.
———. 1994. "The Emergence of Intelligence." *Scientific American*.
———, and George A. Ojeman. 1994. *Conversations with Neil's Brain: The Neural Nature of Thought and Language*. Boston, MA: Addison Wesley.
Campbell, J. 1982. *Grammatical Man*. New York: Simon & Schuster.
———. 1986. *Winston Churchill's Afternoon Nap: A Wide-Awake Inquiry into the Human Nature of Time*. New York: Simon & Schuster.
Cartailhac, Émile. 1902. "La grotte d'Altamira, Espagne. Mea culpa d'un sceptique." *L'Anthropologie* 13:348–354.
Cavalli-Sforza, L.L. and M.W. Feldman. 1983. "Paradox of the Evolution of Communication and of Social Interactivity." *Proceedings of the National Academy of Sciences* 80:2017–2021.
Cheney, Dortothy L. and R.M. Seyfarth. 1990. *How Monkeys See the World: Inside the Mind of Another Species*. Chicago: University of Chicago Press.
Chomsky, N. 1959. Review of *Verbal Behavior*, by B. F. Skinner. *Language* 35:26–58.
———. 1965. *Aspects of the Theory of Syntax*. Cambridge, MA: MIT Press.
———. 1972. *Language and Mind*. New York: Harcourt Brace Jovanovich.
———. 1988. *Language and Problems of Knowledge: The Managua Lectures*. Cambridge, MA: MIT Press.
———. 1991. "Linguistics and Cognitive Science: Problems and Mysteries," in Kasher, Asa, (ed.) 1991. *The Chomskyan Turn*. Oxford: Basil Blackwell.
Crombach, H. F. M., and H. L. G. J. Merkelbach. 1996. *Hervonden herinneringen en andere misverstanden*. Amsterdam, the Netherlands: Contact.
Cunliffe, Barry, ed. 1998. *Prehistoric Europe: An Illustrated History*. Oxford: Oxford University Press.
Damasio, Antonio. 1999. *The Feeling of What Happens: Body and Emotion in the Making of Consciousness*. San Diego, CA: Harcourt Brace.
Darwin, Charles. 1979. *The Origin of Species, Complete and Fully Illustrated*. New York: Gramercy Books.
Dawkins, Richard. 1986. *The Blind Watchmaker*. Longman Scientific & Technical.
Deacon, T. 1997. *The Symbolic Species: The Co-Evolution of Language and the Human Brain*. Harmondsworth: Penguin.
Dennet, Daniel C. 1991. *Consciousness Explained*. New York: Little, Brown & Co.
Diamond, Jared. 1997. *Guns, Germs and Steel: A Short History of Everybody for the Last 13,000 Years*. London: Chatto & Windus.
Dibdin, Michael. 2007. *End Game*. London: Faber & Faber.
Dunbar, Robin. 1993. "Coevolution of Neocortical Size, Group Size and Language in Humans." *Behavioral and Brain Sciences* 16(4): 681–694.
———. 1996. *Grooming, Gossip and the Evolution of Language*. London: Faber & Faber.
Eccles, John C. 1989. *Evolution of the Brain: Creation of the Self*. London: Routledge.
Falk, Dean. 1992. *Braindance: New Discoveries about Human Origins and Brain Evolution*. New York: Henry Holt and Company.

———. 2007. "Constraints on Brain Size: The Radiator Hypothesis," in J. H. Kaas, ed., *The Evolution of Primate Nervous Systems*. Oxford: Academic Press, 347–354.

Fitch, W. Tecumseh, Marc D. Hauser, and Noam Chomsky. 2005. "The Evolution of the Language Faculty: Clarifications and Implications." *Cognition* 97:179–210.

Flatow, Ira. 1992. *They All Laughed*. New York: Harper Collins.

Geist, Valerius. 2000. "The Neanderthal Paradigm." cogweb.ucla.edu/ep/NeanderthalParadigm.html.

Gould, James L., and Carol G. Gould. 1994. *The Animal Mind*. Scientific American Library

———, and Carol G. Gould. 1995. *The Honey Bee*. Scientific American Library.

Greenspan, Stanley I., and Stuart G. Shanker. 2004. *The First Idea: How Symbols, Language, and Intelligence Evolved from Our Primate Ancestors to Modern Humans*. Boston, MA: Da Capo Press.

Guthrie, R. Dale. 2005. *The Nature of Paleolithic Art*. Chicago: University of Chicago Press.

Hadingham, Evan. 1980. *Secrets of the Ice Age: The World of the Cave Artists*. London: Heinemann.

Hahn, J. 1993. "Aurignacian Art in Central Europe," in H. Knecht, A. Pike-Tay and R. White, eds. *Before Lascaux: The Complex Record of the Early Upper Palaeolithic*, Boca Raton, FL: CRC Press, 229–257.

Hamilton, Marcus J., Bruce T. Milne, Robert S. Walker and James H. Brown. 2007. Nonlinear Scaling of Space Use in Human Hunter-Gatherers. *Proceedings of the National Academy of Sciences* 104: 4765–4769.

Haub, Carl .2002. *How Many People Have Ever Lived on Earth*. Population Reference Bureau.

Hauser, Mark D. 1996. *The Evolution of Communication*. Cambridge, MA: MIT Press.

———, Noam Chomsky, and W. Tecumseh Fitch. 2002. "The Faculty of Language: What Is It, Who Has It, and How Did It Evolve?" *Science* 298: 1569–1579.

Hespos, Susan J., and E. S. Spelke. 2004. "Conceptual Precursors to Language." *Nature* 430: 453–456.

Hirsch, Eli. 1982. *The Concept of Identity*. Oxford: Oxford University Press.

Hodges, Andrew. 1983. *Alan Turing: The Enigma of Intelligence*. London: Burnett Books.

Hodder, Ian. 1990. *The Domestication of Europe: Structure and Contingency in Neolithic Societies*. Oxford: Basil Blackwell.

Hofstadter, Douglas R., and Daniel C. Dennett. 1982. *The Mind's I: Fantasies and Reflections on Self and Soul*. London: Penguin.

Homeros, 1959. *Ilias and Odyssea*, tekstgetrouwe weergave door F. van Oldenburg Ermke, Kempische boekhandel.

Janik, V. M., L. S. Sayigh, and R. S. Wells. 2006. "Signature Whistle Shape Conveys Identity Information to Bottlenose Dolphins." *Proceedings of the National Academy of Sciences* 103(21): 8293–8297.

Johnson-Laird, Philip. 1988. *The Computer and the Mind: An Introduction to Cognitive Science*. Cambridge, MA: Harvard University Press.

Kenneally, Christine. 2007. *The First Word: The Search for the Origins of Language*. New York: Viking.

Koenen, Liesbeth. 2000. Zeeslakjestrainer prof. Eric Kandel: 'Door je geheugenonderdrukkingsgenen stop je niet alles wat je tegenkomt in je geheugen'. *Akademie Nieuws* 59.

———. 2008. Doof en blind. *NRC Handelsblad* 2-8-2008.

———, Tony Bloem, Ruud Janssen, and Albert van de Ven. 1998. *Gebarentaal, de taal van doven in Nederland, 3e herziene druk*. Amsterdam, the Netherlands: Atlas.

Lamme, Victor. 2010, *De vrije wil bestaat niet; over wie er echt de baas is in het brein*. Amsterdam, the Netherlands: Bert BakkerLangmore, N. E, G. Maurer, G. J. Adcock, and R. M. Kilner. 2008. "Socially Acquired Host-Specific Mimicry and the Evolution of Host Races in Horsfield's Bronze-Cuckoo Chalcites Basalis." *Evolution*, June 2008: 1689–1699.

Lawick-Goodall, Jane van. 1973. "Cultural Elements in a Chimpanzee Community," in E. Menzel Jr., ed. *Precultural Primate Behavior*. Freiburg: Karger.

Leakey, Richard, and Roger Lewin. 1992. *Origins Reconsidered: In Search of What Makes Us Human*. Abacus.

Leeuwen, E. J. C. van. 2015. "Social Learning Dynamics in Chimpanzees: Reflections on Animal Culture." PhD diss., Radboud University Nijmegen, Nijmegen.

———, Katherine A. Cronin and Daniel B.M. Haun 2014. A Group-specific arbitrary tradition in chimpanzees (*Pan troglodytes) Animal Cognition* 17,6: 1421–1425.

Lieberman, Philip. 1984. *The Biology and Evolution of Language*. Cambridge, MA: Harvard University Press.

———. 1991. *Uniquely Human: The Evolution of Speech, Thought, and Selfless Behavior*. Cambridge, MA: Harvard University Press.

———. 1998. *Eve Spoke: Human Language and Human Evolution*. New York: W.W. Norton & Co.

Louwe Kooijmans, L. P. 1980. Het onderzoek van neolithische nederzettingsterreinen in Nederland anno 1979, *Westerheem* 29.

Marmasse, Natalia e.a. 2000. *Numerical Mechanisms and Children's Concept of Numbers*. MIT Media Laboratory Technical Report.

Merleau-Ponty, Maurice. 1973. *Consciousness and the Acquisition of Language*. Evanston, IL: Northwestern University Press.

Meyers, Elaine. 1985 The Hundredth Monkey Revisited; Going back to the original sources puts a new light on this popular story. *In Context*, http://www.context.org/iclib/ic09/myers/.

Mithen, Steven. 1996. *The Prehistory of Mind: the Cognitive Origins of Art and Science*. London/New York: Thames & Hudson.

Nowak, Martin A., Natalia L. Komarova and Partha Niyogi. 2001. "Evolution of Universal Grammar." *Science* 291/5501: 114–118

———, and David Krakauer. 1999. "The Evolution of Language." *Proceedings of the National Academy of Sciences* 96: 8028–8033.

Parker, Sue T., Jonas Langer, and Constance Milbrath. 2004. *Biology and Knowledge Revisited*. Psychology Press.

Petitto, Laura A., and M. S. Seidenberg. 1979. "On the Evidence for Linguistic Abilities in Signing Apes." *Brain and Language* 8: 72–88.

Petkova, Valeria I., and H. H. Ehrsson. 2008. "If I Were You: Perceptual Illusion of Body Swapping." *PLOS ONE* 3(12).

Piatelli Palmarini, Massimo. 1996. *Onvermijdelijke illusies: Hoe beredeneringsfouten ons denken beheersen*. Utrecht: Het Spectrum.

Pinker, Steven. 1994. *The Language Instinct*. New York: William Morrow.

———. 1997. *How the Mind Works*. New York: W.W. Norton & Company.

———, and P. Bloom. 1990. Natural Language and Natural Selection. *Behavioral and Brain Sciences* 13: 707–784.

———, and Ray Jackendoff. 2005. "The Faculty of Language: What's Special about It?" *Cognition* 95: 201–236.

Ramachandran, V. S. 2004. *A Brief Tour of Human Consciousness*. New York: Pi Press.

———, and Sandra Blakeslee. 1998. *Phantoms in the Brain: Probing the Mysteries of the Human Mind*. New York: William Morrow.

Rose, Steven, R.C. Lewontin, and Leon J. Kamin. 1984. *Not in Our Genes: Biology, Ideology and Human Nature*. New York: Pantheon Books.

Rudgley, Richard, 1999. *The Lost Civilizations of the Stone Age*. New York: The Free Press.

Rundus, Aaron S. 2007. "Ground Squirrels Use an Infrared Signal to Deter Rattlesnake Predation." *Proceedings of the National Academy of Sciences* 104/36: 14372–14376.

s.a. 2006, Dolphins 'have their own names', BBC News May 8, http://news.bbc.co.uk/2/hi/uk_news/scotland/edinburgh_and_east/4750471.stm.

Saint-Gérand, Jacques-Philippe, s.a. *Morgan Kavanagh: Condylure oublié en histoire des sciences du langage?*, Université Blaise Pascal, Clermont-Ferrand II.

Schilthuis, A.P. 2007. *Positieve veranderingen voor de vrouw in de wetgeving*, Nederlandse Vereniging voor Vrouwenbelangen, Vrouwenarbeid en Gelijk Staatsburgerschap.

Schouwstra, Marieke. 2008. *The Honeybee Dance: A Language?* Utrecht University OTS unpublished.

Smith, John Maynard. 1993. *The Theory of Evolution*. Cambridge: Canto.

Smits, Rik. 2010. *The Puzzle of Left-handedness*. London: Reaktion Books.

Sorabji, Richard. 1972. *Aristotle on Memory*. London: Duckworth

Squires, Euan. 1990. *Conscious Mind in the Physical World*. Bristol: Adam Hilger.

Sutherland, Stuart. 1992. *Irrationality: Why We Don't Think Straight*. New Brunswick, NJ: Rutgers University Press.

Tattersall, Ian. 1998. *Becoming Human: Evolution and Human Uniqueness*. Oxford: Oxford University Press.

Thomson, J. Arthur, ed., 1922. *The Outline of Science, Vol. 1: A Plain Story Simply Told*. New Rochelle, NY: The Knickerbocker Press.

Vernon M.D. 1966. *Experiments in Visual Perception*. London: Penguin Books.

Vroon, Piet. 1989. *Tranen van de krokodil: Over de te snelle evolutie van onze hersenen.* Utrecht: Ambo.
Waal, Frans de. 1982 *Chimpansee Politics.* London: Jonathan Cape.
———. 1989. *Peacemaking among Primates.* Cambridge, MA: Harvard University Press.
———. 2007. "Bonobos, Left and Right: Primate Politics Heats Up Again as Liberals and Conservatives Spindoctor Science," *eSkeptic* August.
White, Randall. 2003. *Prehistoric Art: The Symbolic Journey of Humankind.* New York: Harry N. Abrams.
Whiten. Andrew. 2007. "Pan African Culture: Memes and Genes in Wild Chimpanzees." *Proceedings of the National Academy of Sciences* 104/45: 17559–17560.
Whitham J. C., M. S. Gerald, and D. Maestripieri. 2007. "Intended Receivers and Functional Significance of Grunt and Girney Vocalizations in Free-Ranging Female Rhesus Macaques." *Ethology* 113: 862–874.
Zeki, Semir 1993 *A Vision of the Brain.* Oxford: Blackwell Scientific Publications.
Zohary, Daniel and Maria Hopf. 1993. *Domestication of Plants in the Old World.* 2nd edition, Oxford: Oxford University Press.

Index

Aborigines, 157–158
abstraction, 152, 157, 177, 179, 188
accidental, ix, 21, 46, 63–65, 74, 125, 156, 158, 160, 178, 192
accidents, 63, 70, 81
Achilles, 194–195
Adam, viii, ix, 128
Adcock, Greg, 158
adolescents, 59, 135
adornment, personal, 108, 120–121, 162, 189
adultery, 39
Agamemnon, 194–195
agriculture, 2, 117, 173, 180–190
aiming, 146
Alex, parrot, 3
alphabet, 175, 177–180
Altamira, 109–111
altruism, 77, 192
American Civil War, 173, 201
Amindilyakwa, 73
amputation, 81–84, 86–87, 131
amygdala, 79
Andrews, Julie, 164
Andrews, Peter, 12
animal communication system, 25, 41–52, 54
animal rights, 13
anonymity, 192–193
anthropomorphizing, 65
ants, 47–48
apes, 7, 9, 13, 38–40, 44, 97, 100–102, 122, 126, 147
  talking, 7–9, 188
aphasia, 86
Aphrodite, 195
Aplysia, 67–68, 78, 95
apocrine glands, 104
Apollo-11 cave, 136
Appel, Karel, 64–66, 200
Arc, Joan of, 52
Aristotle, 98
arithmetic, 72–73, 102, 152, 153, 177
art pour l'art, 136
art, abstract, 200–201
art, visual, 112, 121–122, 124, 127, 160
ASL, 7
Astralagus, 199
asymmetry of hemispheres, 102, 200
aurochs, 114
Australopithecus afarensis, 1, 96
automaton, 68, 93
axon, 99

Babel, tower of, viii
babirusa, 136
Babyshambles, 103
Barnaby, Chief Inspector, 75
beads, 75, 108, 120–121, 162–163
Beatles, the, 53
Becanus, Goropius, ix–x
bees, 35–36, 47, 59
Bégouën brothers, 107
behaviorism, 5–7, 15, 26, 78, 146
Bentham, Jeremy, 202
Benz, Karl, 169–171
Bible, 117, 131
bilingual, 44
binary system, 72, 80
birds, 11, 27, 29–31, 34, 37, 42–43, 59, 66, 100, 103, 120, 122, 147, 149, 186, 192
birdsong, 42, 164
birth, 2, 5, 18, 19, 37, 62, 78, 81, 96–99, 107, 163
bison, 105–107, 109, 111, 113–114, 130
Black Eve, 12, 190

Blatter, Sepp, 53
blind, 55–57, 59–60, 86, 198
boar, 114
Bocelli, Andrea, 56–57
body language, 49
Boniface, 117
Bonnet, Charles, 88–89
bonobos, 7, 10, 38–39, 131–132, 186
boredom, 2, 16, 139–141, 144–145
Borghese, Scipione, 171
Bouton, Georges, 170
bowerbirds, 120
bowing, 49
brain damage, 62, 86, 89
brain stem, 78, 79, 99
brain, viii, 1, 4–5, 14, 17, 18, 25, 56, 60–63, 67, 72–73, 78–79, 81–84, 86, 88–89, 91, 93, 96–103, 108, 123, 125–127, 134, 136, 142, 145–147, 151–156, 158, 163, 198
breeding, 33, 41, 182, 184, 185
Briseis, 194
Brno, 162
Broca's area, 102
Bruni, Carla, 164
burial, 132, 163, 189

Calvin, William, 100–101
Calypso, 196
car, 51, 58, 76, 79, 144, 170–172
carnival, 192
Cartailhac, Émile, 110
Cartesian theater, 89
Cartesian view, 69
Çatal Hüyük, 186
cats, 85, 123, 139
cave painting, 105, 107–114, 117–121, 124, 127, 129–130, 133, 136, 158, 159, 161, 183
cerebellum, 79, 99
chain reaction, 126
chance, 21, 65, 199–200
Charles, Ray, 56
chess, 166
chicken, 139
chimpanzee, 1, 3, 7, 38, 39, 40, 48, 74, 96, 99, 102, 118, 120, 130, 132, 136, 147, 148
Chimpsky, Nim. *See* Nim
China, 114–116, 175, 176, 178
Chinese, 57, 61
Chomsky, Noam, 3–7, 15–17, 26
christianization, 117
Christopher Robin, 76
Circe, 195–196
circus, 138
civil rights, 204
civilization, 92, 114, 117, 119, 120, 183, 189, 190
cliché, 148
clock, 87
  biological, 147
  genetic, 12
cockroaches, 66
code, coding, 11, 19, 21, 25, 30–32, 34–35, 40–41, 45, 56, 128–129, 137, 167, 177, 180
coincidence. *See* accidental
Coldplay, 54
collective, 49, 53, 57, 115, 131, 133–134, 150, 155, 192–194, 197, 205
Colossus, 167
communication, 7, 11, 25–26, 30–35, 39, 41–54, 154, 191
computer, 72, 102, 139, 146, 166–169, 172, 176, 179
concepts, 4, 20, 56–59, 63, 69–79, 89, 125–126, 128, 147, 152, 167, 177, 187, 188, 192, 198
conditioning, operand, 78
Confucius, 115
consciousness, 13–14, 63, 64, 68, 90–91, 94, 125–126, 132, 142, 153, 155, 159
  self-, 13
cooling, 19, 97–98, 170
cortex, 2, 82, 99
cortisol, 125
counting systems, 71–73, 80
cows, 69, 106, 140, 188
creativity, 2, 79, 91, 95, 97, 126
critical period, 17–18, 61
Cro-Magnon, Abri de, 107
Cro-Magnons, 12, 107–109, 111–114, 117–121, 127, 129–136, 158–160, 183, 187, 190, 198, 205
Crompton's mule, 174
crosswords, 54, 139
crows, 137
cuckoos, 37–38
culture, 2, 5, 10, 13, 15–20, 40, 46, 49, 51, 54, 70–71, 83, 92, 102, 107, 111, 113–117, 120, 132, 149, 155, 162–163, 173, 176, 180, 183, 188–190, 193
curses, 51–52, 177, 191
cycles, 147, 151, 189
Cyclops, 197–198

Da Vinci, Leonardo, 124
dactylus, 148
Daimler, Gottlieb, 170
dancing, 15, 42, 103, 115, 147, 192
  of bees, 35–36, 47–48
Darwin, Charles, vii–viii, 19, 21–24, 46, 66, 102, 154
daydream, 141, 165
deafblindness, 55–56
deceit. *See* lying
decimal system, 72, 80
deer, 109, 114, 138, 199
  Père David's, 120
delusions, viii, 134
democracy, 176, 204
Dendrobatidae, 29
denial, 86–87
Dennet, Daniel, 89
Descartes, René, 68
destruction, 19, 89–90, 115–117, 139
Dibdin, Michael, 137
dice, 199–200
diet, 99–100, 113, 130, 184, 185
Diomede, 195
Dion, Albert de, 170
Disraeli, Benjamin, 200
DNA, 12–13, 19, 107108, 158
dogs, 3, 27, 33, 41, 42, 47, 65, 70, 78, 85, 104, 122, 126, 132, 140
Dolni Vestonice, 162–163
dolphins, 10–11, 126, 149
domestication, 7, 33, 79, 189
dots, 127, 159, 189
drawing, 111, 117, 121, 123, 127, 129, 133
dream, American, 6
dreaming, 126, 135, 142
Duck, Donald, 76, 103
Dumbacher, Jack, 29–30
dung heap, 185
Dunlop, John Boyd, 170
duodecimal system, 72
Dutch, 45, 51, 179
Dyonisia, 164
dyslexia, 62

earthenware. *See* pottery
eccrine glands, 104
Eden, Garden of, ix, 128
ego, 130, 132
electronics, 169, 171
elephants, 48, 59, 126, 138, 139
Elssler, Fanny, 160
emancipation, 91, 202, 204–205
emotions, 5, 51, 71, 79, 93, 124, 139, 150, 179, 191, 194
engineering, 20, 22, 64, 169–174
English, ix, 38, 45, 51, 57, 61, 73, 80, 92, 179
ENIAC, 167, 168
enjoying the view, 144–145
enlightenment, 156, 199
Erdoğan, Recep Tayyip, 204
erotic, 83, 113, 142, 149, 160, 162, 164, 189
Esperanto, ix
ethology, 3, 10
evolution, vii, 12, 20–23, 25, 32, 41, 46, 48, 54, 64–66, 71, 95–96, 98, 101, 108, 122, 124–125, 145–146, 149, 153–154, 155, 163, 190, 199, 202
evolutionary advantage, 25, 26, 46, 95, 98, 101, 122, 146, 149, 153–154, 163
excrement/excretion, 42, 47, 70, 71
expressionism, 200

fabric, 162
face, 20, 41, 49, 83–84, 87, 160, 189
Falk, Dean, 97
Fanny, 160–161
fantasy, 76, 126, 133
farm animals, 182, 189
farming, 69, 139, 181, 183–184
fear, 31, 70–71, 78, 93, 124, 136, 140, 142, 143
  of heights, 69, 71
feedback, 82, 88–89, 91, 126, 136
female sign, 112, 113
Fertile Crescent, 181–183, 185, 189–190
fertility, 18, 105, 106, 160
fertilization, 12, 185
figurines, 120, 159–162, 198
film, 34, 39, 62, 89, 112, 124, 128, 148
fire, 70, 106, 144, 150, 156, 162, 192
fish, 29, 33, 40, 130, 184
flagellum, 64, 67
flies, 66, 185
food, 9, 35, 37, 38, 40, 52, 78, 99–100, 113, 130, 140, 147, 184–186, 196
foot
  erotic meaning of, 81
  metrical, 148
football. *See* soccer
footprint, 1, 105–106, 111
foramina, 97–98
Ford, Henry, 172

fox, arctic, 162, 163
free will, 68, 91–95
freezing as a defense, 123
French, 45, 57, 179
frogs, 29, 123, 124
funeral. *See* burial
fur, 42, 46–47, 98, 104, 113
futhark, 180

Galgenberg Venus, 160–161
gambling, 199–200
Gardner & Gardner, 7
genetic distance, 12
genetic drift, 23, 25
genetic variation, 40, 117
genitals, 81–83
German, 179
girney, 10
goats, 69, 177
God, 51, 76, 128, 131
godams, les, 52
Gogh, Vincent van, 200
Gollum, 49
Goodall, Jane, 40
gorilla, 3
Gould, Stephen J., 46
grain, 177, 183–188
grammar, 16, 46, 92, 128, 149
Greek, ix, 57
grey matter, 97, 99
group animals, 35, 77
group behavior, 33, 35, 38, 59, 73, 77, 102, 103, 120, 132, 147, 149, 181, 191–192, 203
guinea pigs, 141
Gutenberg, Johannes, 174–175

habit formation, 6, 78
hallucinations, 88–89, 103, 141, 143
Hammurabi, 76
hamsters, 141
handprints, 131, 136, 159
hawk, 137
Healey, Jeff, 56
hearing voices, 103, 134, 141
Helen of Troy, 193
hemisphere, 82–83, 86, 102
Hera, 148, 195
heredity, 5–6, 15, 19, 23, 79, 93–94, 138
Hermes, 195
Herod, viii–ix
hexameter, 148
Hilbert, David, 165
hippocampus, 79
Homer, 56, 148, 155, 193, 197
hominids, 12, 21, 25, 96, 98–100, 102–103, 108, 126–127, 146, 155–157
Homo erectus, 96, 100, 127, 146, 156
Homo habilis, 100, 166
Homo sapiens, 14, 108, 117, 127, 156–157
homunculus problem, 89–90
horses, 74, 104, 109, 114, 138, 139, 147, 188
hunter-gatherers, 100, 107, 118, 162, 181, 183–185
hunting magic, 111, 113, 114
hunting scene, 109, 114, 130
hunting, 34, 99–101, 107, 108, 111, 113, 114, 130, 131, 138

ice age, 108, 109, 119, 157
Icelandic, 179
identity, 64, 115, 133–134, 163, 179, 187–189, 192–193, 196–198
Ignatievka, 189
Iliad, 148, 193–196
illusion, 64, 74, 84–85, 90, 92, 123
  body swapping, 85, 90
  rubber hand, 84–85, 90
  visual, 123
imaginary friend, 76, 134
individuality, 115, 130–134, 152–154, 156, 159, 161–163, 187–190, 192, 194, 196–197, 205
Indo-European, ix
infants, ix, 6, 69, 70, 127, 135, 181
infarction, 86
infrared radiation, 42, 59
innate, 5, 7, 15, 17, 36–37, 40, 44, 56, 69–71, 73–74, 78–79, 92–93, 135, 137,145
inner speech, 53
inner world, 56–57, 63–64, 67–70, 75–79, 83–87, 89–96, 125–127, 129, 132–136, 141–146, 150, 152–154, 188, 198, 205
inner world products, 91–92, 152, 188
instinct, 34, 68, 74, 122, 124, 199
instruction, explicit, 9, 17, 34, 39, 166, 168
integration, 41, 62, 79, 152–158, 163, 190
intelligence, 16, 38, 40, 64, 91–92, 95–96, 111, 120, 154, 183, 186, 188
interbreeding, 108
interior monologues, 90
Inuit, 92
inventions, 2, 100, 117, 135, 169, 170, 172–174, 176, 177–179, 180–182, 188, 190, 200
IPA-alphabet, 179

Iran, 204
Islamic State, 116
ivory, 120, 162, 163

Janik, Vincent, 11
Janssen, Marleen, 55
Jaynes, Julian, 155
Jeantaud, Charles, 170
Jenatzky, Camille, 171
Jessup, William, 172
Jones, William, ix
justice, 56, 71, 198

Kandel, Eric, 67–68
Kavanagh, Morgan, vii–viii, x
Kay, John, 173
keeping watch, 155, 188
Keller, Helen, 55
Khrushchev, Nikita, 116
Kiefer, Anselm, 200
Kilby, Jack, 169
kittens, 17, 69
knucklebones, 199–200
Koko, 3
Kooning, Willem de, 200
Kostenki, 189

Lady Gaga, 164
Laetoli, 1
Lake Mungo, 157
lambs, 69
language
  animal, 26
  body, 49
  original, ix
  sign, *See* sign language
  universal, vii, viii
language acquisition, 9, 15, 17–18, 44–46, 73, 128–129, 153, 187
language learning, 3, 7, 9, 14–15, 17, 18, 44–46, 58, 61, 70, 76–77, 175, 179, 187
language organ, 4, 26
Lao-tze, 115
Lascaux, 109, 114, 130
lead type, 175
Leander, Zarah, 164
learning, 3, 6, 7, 9, 14–15, 17, 18, 21, 34–36, 39–40, 44–46, 55, 58, 61, 64, 67–69, 70, 76–78, 95, 121, 125, 134, 135, 138, 145, 148, 175, 179, 185, 187
Lenin, Vladimir, 116
Lenoir, Étienne, 169
limbic system, 79, 98
linguistics, generative, 3
lions, 34
lip-reading, 79
Little Red Riding Hood, 164
logic, ix, 4, 6, 13, 92, 147, 156, 166, 187–188, 197, 198
Lorenz, Konrad, 10
lottery, 65
Louis Vuitton, 116
LSD, 142
Lucy, 1–2, 96, 97
lying, 13, 37, 38, 52, 87, 134
lyrics, 54

macaques, 10, 84, 102–103
mammoths, 113, 162, 163
Mandela, Nelson, 53
marching, 150
marmosets, 141
Maros, 136
masturbation, 38
mathematics, 80, 92, 156, 165–166, 178
meditation, 142–143
meerkats, 33
memory/memories, 5, 13, 16–17, 20, 56, 63–64, 67–68, 78–79, 85, 88, 89, 91, 94, 95–96, 115, 117, 125–126, 134, 142, 145–146, 158, 168, 188, 193
Menelaus, 193–194
menstruation, 147, 160, 189
mice, 99, 122, 186
Michelangelo, 128
middle ages, 175, 176
Milne, A.A., 76
mimicry, 38
mind, 5–6, 14, 18, 39, 57–59, 69, 71–77, 81, 85–86, 119, 124, 127, 129–136, 139, 141–143, 145, 152, 155–157, 184, 188–190, 191, 198–201, 203, 205
  theory of, 39, 133
mind's eye, 89–90, 127, 136
mirror image, 85, 122–123
Mithen, Stephen, 155
mitochondria, 12
model, 4, 13, 22, 31, 63, 172, 174
  sender-receiver, 31, 34–35
Mohammed, 116
Mona Lisa, 124
monkeys, 10, 69
Monroe, Marilyn, 164
moon, 70, 74, 122

moon cycle, 147
mosquitoes, 120
mother tongue, 5, 34, 45, 56, 128, 135
motherese, 10
motherhood, 160
motor skills, 20, 69, 83, 93, 122, 128
Mount Toba, 117
Muhammed, 116
multilingual, 15
Mungo Man, 157–158
music, 53, 55, 61, 103, 138, 139, 149, 153
mutations, 12, 23, 25, 46, 66–67, 77
mutilation, 131
myths, vii, viii, 12, 56, 66, 158, 193, 195, 197

naming, 9, 128, 187–188
Napoleon, 31, 137
nature-nurture, 5
Nazis, 6
Neanderthals, 96, 107–108, 113, 121, 132, 156
neglect, 86–87, 89–90
Nelson, Horatio, 88
neocortex, 2, 82
nerves, nerve endings, 60, 83–84, 88
Neumann, John von, 168
neuroma, 84
neurons, 63, 67–68, 95, 99, 101, 125–126, 146
Niaux, Grotte de, 120
Nim, 3, 7
nobody, 198–199
nomadic, 183–184, 186
nonsense rhymes, 53
nouns, 16, 52
Noyce, Robert, 169
numbers, 71–74, 80, 152–153, 165, 199

oaths. *See* curses
Odyssey, 56, 148, 193–197, 199
olfactory markers, 33
onomatopoeia, 44
opera, 54
orang-utan, 66
Orbison, Roy, 56
orgasm, 81
ornaments, 120–121, 162–162
Out-of-Africa hypothesis, 12

pain, 6, 29, 31, 59, 70, 71, 81, 83–84, 87–88, 96
Pallas Athena, 195, 197
Paris, 193–194
parrot, 3
Parton, Dolly, 164
passive voice, 18
past tense, 18
Pavlov, Ivan, 78
Peking to Paris rally, 171
Penelope, 197
penguins, 11
perception, 124
  categorical, 60–62
permanence, 187
personality changes, 86
personality, 75, 90, 192
Petitto, Laura-Ann, 8
pets, 3, 59
phallic symbols, 105
phantom limbs, 81, 83–89
phantom pain, 83–84, 87–88
pheromones, 33, 40
phobias, 70–71
Phrygian, ix
pig deer. *See* babirusa
piglets, 69
pigments, 120–121, 131, 198
pigs, 139
Pink Floyd, 165
pitohui, 29–32, 52
pituitary gland, 79
plants, 30, 32, 33, 130, 184–186
  carnivorous, 32
plant communication, 30, 32, 33
playing cards, 139, 200
plumage, 29, 31 , 46
poetry, 148, 177
pointing, 127–129, 131, 153
Polyphemus, 197–198
Pons, Timothy, 84
Popper, Karl, viii
pornography, 162, 164
Portuguese, 179
pottery, 162, 186
printing press, 174–176, 179, 200
probability, 65, 199, 200
pronouns, 18
proper names, 10–11, 27, 77
proprioception, 81–82
Proto-Inod-european, ix
Psamtic, viii–ix
psilocin/psilocybin, 142
psychology, 3, 5–7, 93, 113, 152, 179
psychosis, 135, 142

puppies, 69
puzzle, 54, 123, 139

rabbit, 47, 99–100, 102, 126, 141, 150
radiator hypothesis, 97
railways, 172–173
rainbow, 74
Ramachandran, Vilayanur, 81, 83, 85, 87–89
rape, 205
rationality, 13, 64, 92, 93, 138, 156, 197, 202
rats, 141
rattlesnake, 42, 52
reading, 44, 62, 72, 87, 148, 175
real estate, 183
recursion, 150–153
reference, 45, 57–58, 93, 128, 189
reindeer, 108, 113, 114, 130
religion, viii, ix, 105–106, 111, 113–117, 131–132, 136, 204
Renaissance, 115
revolution, 99, 116, 155, 168, 172, 182, 203
  Cultural, 115
  French, 172, 173
  Industrial, 173–174
  Natufian, 162, 189
rhinoceros, 113, 114
rhythm, 39, 53, 103, 133, 147–150, 152–153
risk, 200
rituals, 42, 49, 53, 103, 105–106, 113–114, 131, 147, 180, 189
Rolling Stones, 54
roulette, 200
Roussef, Dilma, 53

Saddam Hussein, 116
Sadiman, 1
Sanskrit, ix
Sanz de Sautuola, Marcelino, 109–110, 111
Sanz de Sautuola, Maria, 111
Saussure, Ferdinand de, 45
Saussurian arbitrariness, 45
Scandinavian, 179
scent, 32, 33, 42, 44, 47, 104
Scrabble, 54
script. *See* writing system
Scylla, 196
sea hare / sea slug. *See Aplysia*
seafaring, 157–158, 196, 197, 202
seedlings, 186, 188
seeing. *See* vision
selection, 19–20, 23, 26, 46, 48, 67, 74, 95, 102, 125, 184
self-awareness, 89–90, 130, 132–133, 193, 199
self-image, 86–87, 122
senses, 55, 59–60, 62–63, 74, 123
sensors / sensory, 42, 63, 82–84, 89, 91, 94, 125–127, 134, 147
sentences, 5, 9, 16, 20, 41, 43–45, 51–52, 86, 128–129, 149, 152–153
settlements, 2, 117, 163, 181, 183, 189
sex / sexuality, 19–20, 23, 47, 81, 104, 112–113, 160–161, 164, 205
sheep, 197, 198, 199
sign language, 7, 9, 27, 43–44, 62, 188
signature, 131, 159
signature whistle, 11, 27
silence, minute of, 53
Simpson, Homer, 76
singers, 42, 54, 149, 164
singing / songs, 42, 48, 53–54, 103, 133, 144, 147, 149–154, 164, 196
Sirens, 32, 196
Skinner, Burrhus Frederic, 26
slaves / slavery, 194, 202, 205
snake, 29, 42, 70, 71, 125
soccer, 5, 27, 49, 53, 59, 103, 138, 192
Société de Linguistique, vii–viii, x, 3
Socrates, 115
software, 4, 63–64, 167
solitary animals, 35
somatosensory strip, 82–84
songbirds, 11, 34, 37, 42–43, 103, 149
soul, 68, 88, 152
sowing, 184, 186
Spanish, 51, 179
speech, 9, 18, 53, 60–62, 86, 149, 154, 155
speech sounds, 79, 86, 148–149, 163, 177–179
speech therapy, 18
Spencer, Herbert, 23, 24
sports, 17, 23, 138, 192
Spyker, 171
squirrel, 42, 52, 186
Stalin, Josef, 116
statistics, 12, 23, 40, 200
Stephenson, George 172–173
Stevens, Jeroen, 132
stimulus deprivation, 141–143
Stone, Joss, 164
Stone Age, 121
stoned, 143

stopgap, 148
Stringer, Chris, 12
suffragettes, 203
Sullivan, Ann, 55–56
Sumerians, 72, 177
sun, 35, 57, 70, 74, 98
Sunghir, 162–163, 189
sunrise / sunset, 74–75
survival of the fittest, 22–24
sweat, 49, 104, 109
swindlers, 72
syllable, 148, 177
symbols, 9, 13, 35, 56, 80, 92, 94, 105, 111–114, 116–118, 120, 127, 166–167, 177–179, 192
synapses, 67–68, 126, 146
syntax, 149, 153

tabula rasa, 6, 78
Taliban, 116
territory, 13, 33, 42, 47, 82–84, 99, 122, 137, 181
Thetis, 195
thinking, 13–14, 56, 64, 79, 90–91, 126–127, 133–135, 138, 141, 142, 145, 149, 152–155, 188, 198, 204
Thomson, Robert, 170
threatening, 33, 35, 41, 42, 52, 146
throwing, 99, 101–103, 127–128, 133, 146, 150, 152–153
tigers, 43, 53, 78, 113, 114, 124, 138
time, 45, 62, 63, 132, 140–141, 144, 187
Tinbergen, Niko, 10, 122
Tiresias, 56
tools, 2, 13, 39, 40, 97, 101–102, 108, 109, 126, 155, 156, 166, 172, 174, 177, 183, 203
torture, 132, 141, 143
Tourette's symdrome, 52
traditions, vii, 13, 40, 106, 114, 117–121, 176, 181, 189, 204
Trevithick, Richard, 59
tripping, 142
Trojan horse, 194, 197
Tuc d'Audoubert, 105–106, 112
Turing, Alan, 165–167
Turing Machine, 166–167
Turkish, 45, 57, 61, 179
Twain, Marc, 200

Ulysses, 32, 194, 199, 205
uniforms, 192
universal grammar, 46, 92
universal machine, 167
Urdu, 92
urge, 11, 14, 31, 56, 93, 113, 121, 128–129, 136, 138, 140, 145, 153, 162, 187, 192, 194, 197

Venuses, 159–161
verse, 148
vervets, 36–37, 41, 52
Victorian age, 113
Vietnam war, 201
violence, 114, 130, 164, 205
vision, 18, 38, 59, 60, 62, 69, 85, 87, 88, 89, 103, 122–124, 188
visions, 143
visual. *See* vision
vocabulary, 16, 25, 45, 92, 128, 153
vulva, 112, 113

Waal, Frans de, 10, 38
war drum, 103
Warlpiri, 73
warning, 31–35, 41–42, 88
Washoe, 7
Waters, Roger, 165
Watson, John Broadus, 6, 78
whales, 48, 59, 149
Willendorf Venus, 159–160
Willibrord, 117
Winehouse, Amy, 54
Winnie the Pooh, 76
witnesses, unreliable, 63, 134
wolves, 101, 109
Wonder, Stevie, 56
woodblock printing, 174
word formation, 45
World War I, 203
World War II, 6, 109, 165, 167, 203
wrath, 194
writing system, 105, 175, 177, 179
  alphabetic, 175, 177–178
  Chinese, 175
  cuneiform, 177–178
  pictographic, 177–178
  runic, 180
  syllabic, 177–178

Zamenhoff, Ludwik, ix
Zen, Aurelio, 137
Zeus, 195